physics

Jim Breithaupt

TEACH YOURSELF BOOKS

For UK order queries: please contact Bookpoint Ltd, 130 Milton Park, Abingdon, Oxon OX14 4SB. Telephone: (44) 01235 827720. Fax: (44) 01235 400454. Lines are open from 9.00–18.00, Monday to Saturday, with a 24-hour message answering service.
Email address: orders@bookpoint.co.uk

For USA order queries: please contact McGraw-Hill Customer Services, PO Box 545, Blacklick, OH 43004-0545, USA. Telephone: 1-800-722-4726. Fax: 1-614-755-5645.

For Canada order queries: please contact McGraw-Hill Ryerson Ltd, 300 Water St, Whitby, Ontario L1N 9B6, Canada. Telephone: 905 430 5000. Fax: 905 430 5020.

British Library Cataloguing in Publication Data
A catalogue record for this title is available from The British Library.

Library of Congress Catalog Card Number: On file

Published by Hodder Headline Plc, 338 Euston Road, London, NW1 3BH.

Published by NTC/Contemporary Publishing, 4255 West Touhy Avenue, Lincolnwood (Chicago), Illinois 60646–1975 U.S.A.

Typeset by Transet Limited, Coventry, England.
Printed in Great Britain for Hodder & Stoughton Educational, a division of Hodder Headline Plc, 338 Euston Road, London NW1 3BH by Cox & Wyman Ltd, Reading, Berkshire.

Impression number	10 9 8 7 6 5 4 3 2 1
Year	2006 2005 2004 2003 2002

CONTENTS

PREFACE

This book is written for the beginner with a general interest in science who is curious to find out about physics, one of the fundamental branches of science. Physics is a subject with an immense scope covering ideas, discoveries and applications that stretch from nature on the smallest scale deep inside atoms to structures at the edge of the observable Universe. Physics as the study of matter and energy underpins every other branch of science as well as technology and engineering. For example, neuroscientists using magnetic resonance scanners developed by physicists can now image the healthy brain and its activities. Archaeologists can determine with great accuracy the age of ancient materials using radioactive dating techniques developed by physicists. Engineers can use new materials such as liquid crystals to make consumer goods and to provide more efficient and economic buildings and vehicles.

Physics affects us all as very few areas of human activity are unaffected by physics and its developments. For example, mobile phones and the Internet would not be possible without electronic microchips which were invented by physicists in the 1960s. Almost everything we do makes use of past or present work by physicists. The skills developed through the study of physics influence the way people and organizations work far beyond the laboratories used by physicists. For example, the World Wide Web was developed by physicists who needed to communicate scientific information to each other from different locations.

This book aims to take the reader through the main branches of physics, including the key stages in the development of the subject from the earliest known discoveries to the most recent theories and the present frontiers of the subject. Knowledge and understanding are developed sequentially and the reader is therefore advised to work through the book in sequence. Mathematical requirements are kept to a minimum and explained where necessary. Important experiments are described and activities are provided at intervals to reinforce essential knowledge and understanding. Worked examples with solutions are provided and each topic ends with a summary and a set of questions. Answers to numerical questions are provided at the end of the book. A glossary of key physics words is also provided at the end of the book together with a comprehensive index.

Physics is an exciting subject, full of fascinating ideas and applications and with a deep history. I hope this book provides a successful introduction to the subject and generates enthusiasm to delve further into the subject.

Acknowledgements

I would like to thank my family and my colleagues at Wigan and Leigh College for their support in the preparation of the book, particularly my wife, Marie, for secretarial support and cheerful encouragement. I am also grateful to the publishing team at Hodder and Stoughton, in particular Helen Hart who initiated the project, and Catherine Coe who edited and coordinated the production of the book.

INTRODUCTION

Welcome to physics, the subject we all rely on for every step we take and every breath we breathe. This book is intended to introduce you to the essential ideas and rules of physics, presented to make the subject as easy as possible. We will be looking at ideas and experiments that stretched the minds of scientists and philosophers in the past but they were the pioneers and we are not so we don't need to explore all the dead ends in the route to the frontiers of the subject. At times, we will need to pause to take in important ideas and facts needed to build up an overview of the subject and to progress further towards the frontiers. We will also need to look at the methods and skills used in scientific investigations and how ideas were developed from investigations and then tested and used. So that's a few clues about how we will be making the journey that lies ahead. Now let's see what the route will cover.

About physics

Physicists tend to be a bit unconventional because physics is a very varied subject and physicists like to be able to turn their knowledge and skills to almost any situation. Richard Feynman (1920–1986) was one of the most creative physicists of the twentieth century, making his name through developing a new understanding of the force between charged objects after wartime duties on the Manhatten Project which created the atom bomb. After the 1986 Challenger space shuttle disaster, Feynman give a simple yet historic demonstration to crash investigators in which he showed clearly that the disaster was caused when an O-shaped sealing ring on the fuel line cracked when it was cooled to the very low temperature associated with the type of fuel used. So physicists get to grips with how and why objects and materials behave as they do and why objects interact and move as they do. In short, physics is about matter, which is what material objects are made from, and it is about energy, which is to do with how objects interact and move as they do. However, to reduce physics to the study of matter and energy is to undervalue the subject as physicists can be found applying their skills and knowledge in a vast range of situations beyond the physics laboratory, for example in archaeology, communications, hospitals, the stock markets, sport and leisure, in fact just about everywhere. So-called 'rocket scientists' are paid large salaries to use their skills of theoretical physics to model and predict the movement of derivatives on the stock markets. The latest generation of body scanners were developed from discoveries by physicists working on a phenomenon known as magnetic resonance. These scanners have revolutionized the study of the brain, making it possible for scientists to see how different parts of the brain respond to different experiences and situations.

The impact of physics

We live in the Scientific Age which has brought immense benefits to us all. Compare your lifestyle with that of someone two centuries ago and you will realize that the vast majority of people then lived in abject poverty, dependent for their very lives on the vagaries of the weather which could ruin harvests and hit transportation of food and other essential commodities. The Scientific Age has brought power on demand so we need not rely on wood or coal for cooking and heating our homes. Travel is much quicker, cheaper and far more convenient now than two centuries ago when sea journeys took weeks and overland journeys days. People are much healthier, fitter and live longer now than then, thanks to improvements such as clean water, better nutrition and medical drugs. All these improvements have come about because of discoveries made by scientists. To appreciate the impact of physics, consider just two discoveries made in physics in the nineteenth century which literally revolutionized life in the twentieth century.

Michael Faraday (1791–1867) discovered how to generate electricity and thereby laid the foundations for the electrical supply industry and for electrical communications. Faraday gained an appetite for science in his first job as a bookbinder's apprentice. As well as binding the books, he read them avidly and learned about electricity from them. He wrote to Sir Humphry Davy at the Royal Institution in London and gained a position as an assistant as a result. His research into electromagnetism led to his famous discovery of how to generate electricity using magnets. He developed great insight into the link between electricity and magnetism and discovered the fundamental principles that led to the invention of the alternating current generator, the transformer, the electric motor and the electric telegraph. The electricity supply industry developed in the latter half of the nineteenth century from Faraday's discoveries. By the first decades of the twentieth century, most homes in Britain and other industrial nations had been wired up to a local electricity supplier and nationwide grid systems for electricity distribution were established. Nowadays, we take our electricity supplies for granted unless a power cut occurs and we are then thrown temporarily back into the pre-electricity era. When Faraday demonstrated his discoveries to an invited audience at the Royal Institution, he was asked the question 'What use is electricity, Mr Faraday?' He replied with another question, 'What use is a new baby?' No one can tell what can grow from a new discovery.

James Clerk Maxwell (1831–1879) was a mathematical physicist who used Faraday's ideas to devise a theory that light is an electromagnetic wave. He showed an electric wave and a magnetic wave can travel together through space at a speed of 300 000 kilometres per second, the same as the speed of light. He therefore deduced that light must be an electromagnetic wave. He already knew that the spectrum of light covers a continuous range of colours; namely red, orange, yellow, green, blue and violet. His theory showed that all the colours of light are electromagnetic waves of different wavelengths. When he put forward this theory in 1865, he knew that infrared radiation lies just beyond the red part of the visible spectrum and that ultraviolet radiation lies just beyond the violet part of the spectrum. His theory of electromagnetic waves thus accounted for infrared and ultraviolet radiation as well as light. However, he predicted the existence of electromagnetic radiation outside the known spectrum of infrared radiation, light and ultraviolet radiation. Two decades later, Heinrich Hertz discovered radio waves and he was able to show by measurement that they travel at the same speed as light, thus providing clear evidence that radio waves are electromagnetic waves. Hertz's discoveries were taken up by other scientists including Marconi who led the way forward into the use of radio waves in wireless communications. The enormous range of the radio spectrum is now used for many purposes including radio and TV broadcasting, mobile phone communications, radar systems, satellite communications and navigation. Maxwell's theory of electromagnetic waves provides the theoretical basis for all these forms of communications.

Many more examples can be found of the way in which fundamental discoveries in physics have dramatically altered the way we live. The discovery of the atomic nucleus by Ernest Rutherford and the invention of the transistor are two examples of experimental discoveries in physics that have led to the growth of new industries. The theories of relativity by Albert Einstein and the development of quantum mechanics by Niels Bohr and others are two examples of theories that overturned established ideas of physics and provided radical new approaches in the subject. We will meet these and other shattering new discoveries and ideas later on in this book. However, do not imagine such dramatic changes are all history now. Physicists continue to produce astonishing new discoveries and amazing new ideas. Antimatter, black holes, superstrings are but a few of the latest hot topics in physics. The explosive growth of the World Wide Web is another result of physicists at work barely a decade ago, in this case developing an electronic method of keeping in touch with each other and then taken up at an astounding pace by people and businesses in all walks of life.

The golden rule of science

The methods used by physicists follow a general pattern which is known as the *scientific method*. The key feature of the scientific method is that scientific theories and laws can never be proved, only disproved. No amount of experimental evidence can prove a scientific theory or law yet just one reliable experiment is sufficient to disprove any scientific theory or law. Forget this golden rule of science and disaster is inevitable. In the 1990s, concerns arose that there could be a link between BSE in cattle in Britain and human deaths due to CJD. Other EU countries banned imports of British beef. Top medical and food scientists were unable at the time to prove BSE and CJD deaths were linked. However, the absence of proof does not mean that the link does not exist. They forgot the golden rule that science works by disproving theories and predictions. Absence of proof of the link is not the same as disproving the link. Also, the discovery of a different cause of CJD would not disprove that BSE and CJD are linked because there might be more than one cause of CJD. This terrible tragedy might have been averted if the golden rule of science had not been forgotten.

The scientific method uses experiments and observations to test the laws and theories of physics. A scientific theory or law starts out as a hypothesis, which is no more than a proposal based on the available scientific evidence. Thus a hypothesis about how life started on Earth might be that life in the form of microorganisms was brought to Earth by some means from Mars. This particular hypothesis was put forward as a result of the discovery of microbe fossils in a meteorite found in Antarctica that is thought to have come from Mars. The hypothesis has been tested by robot vehicles landed on Mars which have searched for signs of life in the martian soil and atmosphere. No positive findings have yet been reported so the hypothesis remains just that. However, the discovery of microbes, past or present, on Mars would boost the status of the hypothesis sufficiently to make it into a theory. Further space missions to Mars are being planned and evidence for life on Mars, past or present, may well be found. The theory that life on Earth arrived from Mars would be strengthened if microbe fossils were found on Mars similar to the ones found in Antarctica. The scientific evidence in favour of the theory would then be very strong but it would not be possible to say the theory is true because life on Earth might have developed before the arrival of the microbe fossil from Mars arrived. No amount of experimentation could ever prove the theory but successive tests could reinforce it and boost its status.

The theory of continental drift developed from a hypothesis put forward by Alfred Wegener in 1915. He realized that the similarities between the eastern coastline of South America and the western coastline of Africa could mean they were once joined together and have drifted apart since. He discovered many similarities in fossils, plants and animals on both continents so he put forward the hypothesis that the continents were once all joined as a single land mass. Wegener was unable to find a convincing source of power to make the continents drift apart but his ideas were taken up by Arthur Holmes in 1931 who showed that heat released inside the Earth due to radioactive decay could provide the necessary power. Further observations and measurements over several decades showed that the outer layer of the Earth is made up of separate pieces which fit together to make a sphere. These pieces, known as tectonic plates, carry the continents which are known to drift across the surface as the plates gradually move. Earthquakes and volcanoes occur where the plates are in collision. Wegener's hypothesis thus developed into the theory of plate tectonics as more and more scientific evidence accumulated in support of the ideas. More and more scientific evidence in support of the theory would reinforce its status even more. However, the theory would need to be discarded or drastically altered if reliable scientific evidence against it was ever found. No theory can ever be proved as a single reliable experiment in future could disprove it.

Scientific blunders

The history of science has many cul-de-sacs where discarded theories lie abandoned and forgotten. Some of these theories were once held aloft as part of the litany of scientific wisdom. For example, over two centuries ago, scientists believed that heat is a fluid that flows from hot objects to cold objects, like water flowing along a river. The fluid was called 'caloric' and it gave rise to the calorie as the unit of heat. The supporters of the caloric theory were undaunted when they were asked to explain where the heat comes from when you rub your hands together. They imagined that tiny particles created by the rubbing action released caloric in the process. The caloric theory was opposed by scientists who reckoned that heat is a form of energy. The caloric theory was disproved in 1799 by Sir Humphry Davy. He showed that when two blocks of ice below freezing point were rubbed together, water was produced as a result of the ice melting. He demonstrated that the melting of ice occurs because of friction between the two surfaces in contact. Energy must be supplied to the blocks to make them slide over each other and this energy causes the ice to melt. No mention was needed of caloric nor of particles rubbed off the ice.

Cold fusion is a more recent example of a scientific theory that bit the dust. When the nuclei of two light atoms are fused together, energy is released because the two nuclei bind together to form a single nucleus. This process takes place in the core of a star and is responsible for the energy radiated from the star. The immense pressure at the centre of a star forces light nuclei together and makes them fuse. Over the past four decades, scientists have tried without success to make a fusion reactor which would produce energy on a large scale at a steady rate. The particles to be fused need to collide at enormous speed to fuse together but at such speeds they are very difficult to control. Two decades ago, a breakthrough was claimed by scientists working on an entirely different approach to the fusion problem. They reckoned they had discovered fusion taking place in the spaces between the surface atoms of a certain metal when light atoms were in contact with the metal under certain conditions. The process was referred to as cold fusion because it seemingly did not require the enormous temperatures necessary for fusion in stars. Other scientists disputed that the release of energy was due to fusion and attempted without success to reproduce the process of cold fusion. They concluded that the energy released was not due to fusion and the claim was rejected. A successful fusion reactor, hot or cold, would be a major step forward in providing energy without pollution and without producing greenhouse gases such as carbon dioxide. Cold fusion proved to be a claim too far.

Strange ideas

Do not be too harsh on scientists whose claims are rejected. In 1905, Albert Einstein produced two scientific papers that revolutionized physics and changed our understanding of nature dramatically. One of these two papers was about the nature of light which Einstein showed to be in the form of packets of energy he called photons. His other paper, his theory of relativity, proved that absolute space and absolute time do not exist and that mass and energy are linked. We will meet these ideas later. Both papers were revolutionary yet Einstein was to wait over 15 years before he was awarded the Nobel prize for physics. Why such a long time? The prize is awarded for discoveries that benefit the human race. More than a decade passed before sufficient evidence was available to support Einstein's theories of relativity and light. In recent years, experimental evidence has been discovered for the amazing theories about black holes developed by theoreticians such as Stephen Hawking. These strange objects are cosmic crushers from which nothing, not even light, can escape. A black hole would be likely to destroy human life as well as our planet. Maybe there has been no Nobel prize for Hawking yet as so far knowledge of black holes seems to have brought few benefits. However, you could argue that knowledge that helps us to avoid a black hole would definitely be a benefit to the human race! We will return to the subject of black holes later in this book.

Science in general and physics in particular is about understanding nature from the smallest scale deep inside the atom to the largest scale stretching to the edge of the Universe. The general method used in science is to look for patterns in observations and experimental data then formulate a hypothesis about the patterns or data. The hypothesis is then used to make a prediction which can be tested by appropriate experiments. If the results of the experiment confirm the prediction, the hypothesis is then used to form a theory which is tested by more experiments. If the results do not confirm the prediction, the hypothesis must be modified or abandoned. The more experiments that support the theory, the stronger the theory becomes. The Principle of Conservation of Energy is a theory that has withstood many tests. The Principle states that energy is always conserved in any change. The total energy after a change is the same as the total energy before the change. This key principle did not seem to work in a radioactive process known as beta emission. Energy seemed to disappear without trace in this process. It was concluded that either the Principle of Conservation of Energy does not work in this process or a particle that could not be detected carries away the missing energy. This elusive particle was given a name, the neutrino. Twenty years after the problem was raised, neutrinos were at last detected and the Principle of Conservation of Energy was saved.

Physics is about probing the laws of nature and challenging the status quo. This book is intended to explain the key ideas of physics at the present time but be in no doubt that new and strange ideas lie ahead. More science has been discovered in the last half century than ever before. Physics is an exciting subject for anyone who thrives on changes and challenges.

1 STARTING PHYSICS

Physics has always been at the leading edge of human thought, not just in recent times but also before the Scientific Age which began about four centuries ago. The theories of science that were held to be true long ago might seem very odd when we compare them with our present knowledge. We mustn't forget though that the scientists of ancient times or natural philosophers as they were then called came up with ideas and theories that were astonishingly sophisticated in comparison with other aspects of life in those times. In this chapter we will look at some of the physics ideas from long ago to see just how these ideas were developed and used. We will then move on to look at some important skills needed in physics before moving on to use these skills to carry out density tests.

Physics in practice

Before the Scientific Age

Many theories of science before the Scientific Age were based on the assumption that the Earth is at the centre of the Universe and that living beings were created by one or more superior beings who designated a special role for humans. The idea that humans evolved from apes over thousands of centuries found little favour as it omitted the role of a creator. Theories about the natural world were usually chosen on grounds we would consider unscientific and selected facts were used to support the theories. Other facts that did not match the theories were discarded as unreliable or imperfect. Not surprisingly, alchemy and astrology were two major strands of scientific endeavour before the Scientific Age. For example, attempts to turn lead into gold or to predict events occupied the working lives of many individuals, undoubtedly financed by rich and powerful patrons who imagined that they would amass further wealth as a result of such activities.

Science as a recorded activity flourished in the culture of Ancient Greece and the Mediterranean civilizations which developed from Greece. The idea that matter is composed of atoms was a theory put forward by Democritus (470–400 BC). Two centuries later, Aristarchus put forward the theory that the Sun was at the centre of the Universe. In the next century, Archimedes (287–212 BC) made important discoveries in mechanics and mathematics. The importance of the scientific heritage bequeathed by these and other natural philosophers of the ancient Mediterranean civilizations is undisputed. Indeed the impact of one particular natural philosopher, Aristotle (382–322 BC), was to influence the conduct of science for many centuries right through to the beginning of the Scientific Age.

Aristotle was trained by the philosopher Plato who founded the Academy in Athens. Aristotle accepted Plato's theory of ideas as 'eternal patterns' that lie beyond the natural world. However, he recognized the importance of observations in formulating ideas about nature. Where observations could not be used to decide between competing theories, Aristotle rejected any theory that did not support the accepted ideals and overall philosophy developed by Plato. For example, Aristotle rejected Aristarchus' theory that the Sun not the Earth was at the centre of the Universe. There was no observational evidence that the Earth was rushing through space round the Sun and no evidence either to support the associated idea that the Earth is spinning. Nor would Aristotle accept the idea that matter is composed of atoms. He rejected the idea of atoms in favour of the theory that all matter is made of the four elements; earth, water, air and fire. This older theory fitted the idea that the Earth is at the centre of the Universe, water in the seas lies above the Earth, air lies above the seas and fire is in the heavens.

Aristotle shaped science into a coherent set of ideas that were consistent with the prevailing world view, namely that the Earth is the centre of the Universe and therefore a special place has been accorded to the human race by the creator of the Universe. However, Aristotle's rejection of theories that did not fit in meant that interesting ideas were not followed up and investigated further.

Aristotle stamped his method on science so firmly that it was to last for over 15 centuries. His approach of picking facts to support accepted theories dominated the way science was conducted long after his death, from Ancient Greece, through the Roman Empire, the Dark Ages and into the Middle Ages. Perhaps Aristotle's authority enabled science to survive over this long period, especially through the Dark Ages. The Church promoted Aristotle's scientific method, in particular the model of the Universe developed by Ptolemy a century after Aristotle. Ptolemy held that the Earth was at the centre of the Universe with the Sun and Moon moving round the Earth in different circular orbits. Each planet moved round on a circle whose centre moved round on its own circular orbit. This model explained the movement of the Sun, Moon and planets and fitted in with the concept of the 'geocentric' universe. We will look later in this chapter at the bitter struggle between Galileo and the Church which eventually led to the overthrow of Ptolemy's model.

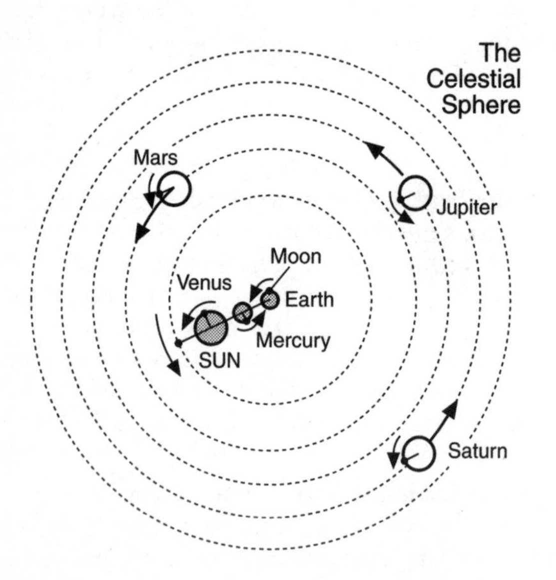

Figure 1A Ptolemy's model of the Solar System

The five base units of the S.I. system are:

1 *the second* which is defined as the time taken for a specified number of vibrations of a certain type of atom in an atomic clock;
2 *the metre* which is defined as the distance travelled by light in a vacuum in a specified time;
3 *the kilogram* which is the unit of mass. This is defined as the quantity of matter in a block of platinum kept in the Bureau of International Weights and Measures (BIPM) in Paris. Standard masses kept in other scientific laboratories are measured by comparison with the standard one kilogram mass in Paris. The weight of an object is the force of gravity on the object. Because the weight of an object is in proportion to its mass, a beam balance as described earlier may be used to compare the masses of any two objects.
4 *the kelvin* is the scientific unit of temperature. See p. 63.
5 *the ampere* is the scientific unit of electric current. See p. 93.

All other scientific units are derived from these five base units. Some examples of derived units are given in the table below.

Quantity	Definition	Unit
Area of a rectangle	length × breadth	square metres, m^2
Volume of a box	length × breadth × height	cubic metres, m^3
Density	mass per unit volume	kilogram per cubic metre, kg/m^3
Speed	distance moved per unit of time	metre per second, m/s

Powers of ten

Scientific measurements and calculations involve values which can range from extremely small to enormously large. For example, the diameter of an atom is about 0.000 000 000 3 metres and the distance from the Earth to the Sun is about 150 000 000 000 metres. Such values are usually written in *standard form* as a number between 1 and 10 multiplied by an appropriate power of ten. For example, the distance from the Earth to the Sun written in this way is 1.5×10^{11} m, where m is the abbreviation for metres and $10^{11} = 100\,000\,000\,000$. Note that 10^{11} is quoted as 'ten to the power eleven'. The above value for the diameter of an atom would be written as 3.0×10^{-10} m where $10^{-10} = 1 \div 10^{10} = 0.000\,000\,000\,1$. Note that a negative power of ten is a code for expressing powers of ten less than 1. Thus 10^{-10} is 1 divided by 'ten to the power ten', usually quoted as 'ten to the power minus ten'.

Scientific prefixes are used to represent certain powers of ten. For example, the prefix kilo- represents a thousand ($= 10^3$). Other standard scientific prefixes are listed below.

Prefix	Tera	Giga	Mega	kilo	milli	micro	nano	pico
Prefix symbol	T	G	M	k	m	μ	n	p
Power of ten	10^{12}	10^{9}	10^{6}	10^{3}	10^{-3}	10^{-6}	10^{-9}	10^{-12}

Notes:

1 The prefix symbol for micro, μ, is pronounced 'mu'.

2 1 centimetre (cm) = 10^{-2} m and 1000 grams = 1 kilogram.

Questions

Q1. Write the following values in scientific form in the same unit (a) 15 000 m, (b) 0.000 045 m, (c) 650 000 000 kg, (d) 0.000 000 003 5 kg.

Q2. Write the following values in scientific form in the required unit (a) 75 mm in metres, (b) 159 cm in metres, (c) 56 000 km in metres, (d) 65 grams in kilograms (e) 0.027 μg in kilograms.

Q3. (a) How many millimetres are there in a length of 1 metre?
(b) How many milligrams are there in 1 kilogram?

Density tests

Which is heavier, a kilogram of lead or a kilogram of feathers? This is a trick question because the weight of a kilogram of any substance on the Earth is the same, regardless of the substance. Anyone who answers that the lead is heavier than the feathers has made the mistake of thinking in terms of equal volumes of the two substances rather than equal masses. The point is that lead is much more dense than a sack of feathers.

The **density** of a substance is its mass per unit volume. The scientific unit of density is the kilogram per cubic metre (kg/m^3).

For example, the density of lead is about 11 000 kg/m^3. In comparison, the density of water is 1000 kg/m^3. Thus lead is 11 times more dense than water. The mass of a certain volume of lead is 11 times greater than the mass of the same volume of water.

Note: the mass of a substance is a measure of the amount of matter the substance contains. Volume is the amount of space a substance takes up.

Density calculations

$$\text{Density} = \frac{\text{mass}}{\text{volume}}$$

The unit of density is the kilogram per cubic metre (kg/m^3). The above formula can be rearranged as

$$\text{mass} = \text{volume} \times \text{density} \quad \text{or} \quad \text{volume} = \frac{\text{mass}}{\text{density}}$$

Worked example

Calculate the mass of a volume of 20 m^3 of water. The density of water = 1000 kg/m^3

Solution

Mass = volume $\times$ density = 20 $\times$ 1000 = 20 000 kg

Density measurements

1. Liquids

- The volume of a liquid is measured by pouring the liquid into an empty measuring cylinder and measuring the level of the liquid in the cylinder against its graduated scale. The scale is usually marked in cubic centimetres (cm^3). To convert the reading into cubic metres (m^3), divide the measurement in cm^3 by 1 million (= 10^6) because 1 $m^3 = 10^6$ cm^3.

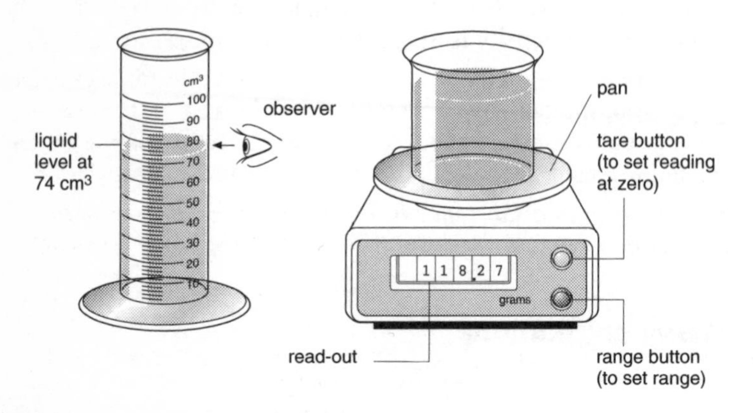

Figure 1C Using a measuring cylinder

- The mass of a measured volume of liquid can be measured by using a top pan balance. This is used to measure the mass of an empty beaker then the liquid is poured into the beaker and the total mass of the beaker and the liquid is remeasured. The mass of the liquid is the difference between the total mass of the beaker and liquid and the mass of the empty beaker. A top pan balance usually gives a reading in grams (g). Note that 1000 g = 1 kg.
- The density is calculated by dividing the mass in kilograms by the volume in m^3.

Sample measurements

Volume of liquid in measuring cylinder = 90 cm^3 = 90×10^{-6} m^3

Mass of empty beaker = 115 g

Mass of beaker and liquid from measuring cylinder = 220 g

$\therefore$ Mass of liquid = 220 – 115 = 105 g = 0.105 kg

Density of liquid = $\dfrac{0.105 \text{ kg}}{90 \times 10^{-6} \text{ m}^3}$ = 1170 kg/m^3

2. Solids

- Use a top pan balance to measure the mass of a piece of the solid.
- Pour some water into a suitable measuring cylinder. Note the volume of the water from the reading of the water level. Tie the piece of solid to a thread and lower it into the water until it is completely submerged. The level of water in the cylinder will rise as a result. Note the new reading of the water level. The volume of the piece of solid is the difference between the two readings.
- The density is calculated by dividing the mass in kilograms by the volume in m^3.

Note: 1. The solid must be insoluble in water.

2. The volume of a regular solid such as a cube can be calculated from its dimensions. For example, the volume of a rectangular box is equal to its length × its height × its width.

Sample measurements

Mass of object = 35.4 g = 0.0354 kg

Volume of water only in measuring cylinder = 50 cm^3

Volume of water and object in measuring cylinder = 63 cm^3

$\therefore$ Volume of object = 63 – 50 = 13 cm^3 = 13×10^{-6} m^3

Density of object = mass / volume = $\dfrac{0.0354 \text{ kg}}{13 \times 10^{-6} \text{ m}^3}$ = 2720 kg/m^3

Eureka!

Perhaps the most famous story in science is about Archimedes when he was asked by his King to find out if his new crown really was made of gold. The King thought the royal crownmaker might have cheated and made cavities in the gold which could have been filled with lead. He didn't want to cut the new crown up so he asked Archimedes to solve the problem without cutting the crown. The solution came to Archimedes when he entered a bath of water and observed that the water level rose as he lowered himself into the water. He realized that when an object is lowered completely into water, the displacement of the water is a measure of the volume of the object. According to legend, he greeted this discovery by running naked through the streets shouting 'Eureka!' which means 'I have found it'.

So how did Archimedes use this discovery to find out if the crown was a fake? Archimedes recognized that the weight of the crown in air divided by its volume should be the same as for any other gold object if the crown was 100% gold. So he weighed the crown and then he measured its volume by measuring the volume of water displaced when it was immersed completely in water. He then calculated the weight of the crown in air divided by its volume. Then he repeated the test and calculation on an object known to be made of gold. Fortunately for the royal crownmaker, the result was the same. Archimedes' measurements showed that the crown's density was the same as the density of a solid gold object. Gold is 18 times more dense than water. In effect, Archimedes discovered that the crown and a known gold object were both 18 times as dense as water. He therefore concluded that the crown was made of gold. Archimedes' test is the first recorded example of non-destructive testing.

Summary

The balance rule

For a beam balanced at its centre of gravity with a weight supported on each side of the beam, the distance of each weight to the fulcrum is in inverse proportion to the amount of weight.

Density

$$\text{Density} = \frac{\text{mass}}{\text{volume}}$$

The unit of density is the kilogram per cubic metre (kg/m^3).

Questions

Q4. Calculate (a) the mass of a volume of 5.0 m^3 of sand of density 2500 kg m^{-3},
(b) the volume of a mass of 40 kg of steel of density 8000 kg/m^3
(c) the density of a brick of mass 6.0 kg and volume 0.002 m^3.

Q5. A wooden cube of mass 0.80 kg measures 0.10 m $\times$ 0.10 m $\times$ 0.10 m. Calculate (a) its volume, (b) its density.

Q6. A glass pane has a height of 0.50 m, a width of 0.40 m and a thickness of 0.006 m.

(a) Show that its volume is 0.0012 m^3.
(b) The density of the glass is 2600 kg/m^3. Calculate the mass of this glass pane.

Q7. Air at atmospheric pressure and at room temperature has a density of 1.2 kg/m^3. Calculate the mass of air in a room of dimensions 5.0 m long $\times$ 4.0 m wide $\times$ 3.0 m high.

Q8. Water has a density of 1000 kg/m^3. Calculate the mass of water in a water tank of volume 6.0 m^3 when the tank is full of water.

Q9. Calculate the density of a liquid from the following measurements.

Volume of liquid = 56 cm^3
Mass of empty beaker = 95.3 g, Mass of beaker and liquid = 147.6 g

Q10. Calculate the density of a solid from the following measurements:

Mass of solid = 48.8 g

Volume of water in measuring cylinder without solid = 40.0 cm^3

Volume of water in measuring cylinder with solid immersed in water = 46.5 cm^3

2 SCIENCE IN MOTION

The Scientific Age in which we live began uneasily several centuries ago. Perhaps the discovery of the New World made some scientists and philosophers think about nature differently to those who accepted Aristotle's teachings. Galileo is generally considered to be the founder of modern science because he demonstrated that Aristotle's way of doing science was fundamentally flawed because it ignored the crucial role of observations. Over succeeding centuries, Galileo's approach to science was developed and the link between laws and observations became evident as new discoveries were made, analysed and tested. The modern philosophy of science is that laws and theories hold on the basis of never having being disproved. This philosophy was formulated by Sir Karl Popper who realized that observations and measurements can never prove a theory or a law but at any time, a single experiment is sufficient to disprove a theory or law.

Measuring motion

Galileo challenged the Church because his astronomical observations led him to conclude that the planets move round the Sun, not round the Earth. This heliocentric model was rediscovered a century before Galileo by Copernicus, a Polish-born medieval scholar whose interests included astronomy, economics, mathematics and medicine. Copernicus devoted many years of his life to studying ancient works to find out why objections to Ptolemy's model of the Universe had been rejected. He constructed his own version of Ptolemy's model and found it was necessary to have more than 30 spheres to account for all the observations of the motion of the stars and the planets. He realized such complexity was not necessary if the Sun not the Earth was at the centre, and he reduced the model to little more than a set of concentric circles representing the orbits of the planets. Copernicus feared that his revolutionary ideas would be ridiculed by his contemporaries and his major work *De Revolutionibus Orbium Coelestium* was not published until shortly before he died in 1543. It was presented as a mathematical challenge rather than a new scientific theory and it failed to have any immediate impact. However, in 1600, Giordano Bruno was burned at the stake by the established Church for promoting the Copernican model and his own view that the Universe is not finite as the Celestial Sphere model implies. Surprisingly, the Copernican model was not fully accepted by the Church until the early nineteenth century. In contrast, the modern Church is far more willing to listen and has organized scientific conferences on current developments in cosmology.

The change from the authoritarian and rigid theories of the medieval age to the experimental philosophy that underpins our Scientific Age may not have originated with Galileo but there is little doubt that Galileo showed through his discoveries the power of science to change established views. Equally important, he showed through his methods the significance of reliable and accurate observations. However, perhaps his most important contribution was to establish that science has limits which should not be extended without valid evidence. In showing that experimental science should be used to explain what we observe, he cut the link between physical science and religion that had existed for centuries. The link continued in biological science well into the nineteenth century through the controversies surrounding Charles Darwin's theory of evolution. Even in this field, the modern science of genetics which supports evolution has been developed through scientific methods developed from Galileo's approach to science.

Galileo's experiments on motion

In a legendary demonstration at Pisa in Italy, Galileo was reported to have demonstrated that two falling objects descend at the same rate, regardless of their weight. This could have been done by releasing two objects simultaneously from the top of the Leaning

Tower and observing that they hit the ground at the same time, thus proving they fall at the same rate.

Galileo wanted to find out how the time of descent would change if the distance fallen was changed. However, he was unable to measure the time of descent accurately but he realized that the motion of an object released at the top of an inclined plane was likely to be similar but slow enough to time. He devised a water clock to measure the time taken by the object to reach certain distances down the plane. Pendulum clocks and other mechanical timing devices were inventions of the future. Galileo measured the amount of water running at a constant rate out of a tank from the moment the object was released at the top of the inclined plane to when it passed a certain marker. The amount of water collected thus served as a measure of the time of descent. You can repeat Galileo's experiment by using a stopwatch rather than a water clock to time a ball rolling down a flat inclined board.

1. Start with zero incline and you will discover that the ball stops rolling after it has been given a push. The reason is that friction between the ball and the inclined plane makes the ball stop.
2. Adjust the incline just enough so the ball rolls along it at steady speed after it is given a push. Record the time taken by the ball to reach markers at equal distances down the incline. The measurements should show the ball travels equal distances in equal times. This is an example of motion at **constant speed**.
3. Increase the incline so the ball rolls down the incline without being given an initial push. Record the time taken by the ball to reach equal distances down the incline. The measurements should show the time the ball takes to pass from one marker to the next decreases as the ball moves down the slope. This is because the ball's speed increases as it moves down the slope. This is an example of **accelerated motion**.

Galileo Galilei

Galileo was born in 1564 in Pisa, Italy, the same year of birth as William Shakespeare. As the son of a nobleman, Galileo was educated in a monastery and in 1595 became Professor of Mathematics at the University of Padua, one of Europe's leading universities at that time, in what was then the Republic of Venice. His Venetian paymasters allowed him to follow his own interests and his discoveries on motion would have been sufficient to win long-lasting recognition. From time to time, he made scientific instruments for commercial purposes such as measuring the density of precious metals and stones. In 1609, reports reached him about the invention of an optical device, the telescope, for making distant objects appear larger and closer. Within a short time, he had designed and constructed his own telescope, capable of making distant objects appear ten times larger and closer. He demonstrated the power of his telescope to the Senate at the top of the Campanile in Venice by observing incoming ships 50 miles away, two or more days' sailing away from the port. Such information was very valuable to the merchants and brokers of Venice.

Galileo realized he could use the telescope to study heavenly bodies. He wanted to gather evidence in support of the Copernican model and he devised a more powerful telescope, capable of magnifying objects 30 times. He was astounded when he first used this telescope to study the night sky. He observed ten times as many stars as can be seen directly without a telescope. He found that the surface of the Moon is heavily cratered and he discovered the four innermost moons of Jupiter. Galileo's astronomical discoveries were widely reported in Europe. His Copernican views became known to the Church who opened a file on him. Galileo hoped his observations and conclusions in support of the Copernican model would be accepted by the Church so he would no longer need to rely on the protection of the Venetian authorities who were anti-clerical. In 1613, he was reprimanded by the Church for his views and three years later was banned from supporting the Copernical model. His attempts over the next decade to persuade the Church to support his views met with failure so he decided to set out his views and his support for the Copernican model in print in Italian, for all to read. He completed his work, *Dialogue on the Two Chief World Systems*, by 1630 and after some difficulties with the censor authorities, it was published in Florence in 1632. Galileo's book was an instant best-seller and the Church reacted rapidly by banning it and stopping further reprints. Galileo, now in his seventieth year, was summoned to appear before the Holy Office of the Inquisition in Rome on April 12, 1633. Galileo duly appeared in front of the 12 judges of the Inquisition. No consideration was given to the discoveries made by Galileo. Instead, the judges sought to show that Galileo had breached the agreement of 1616 when he was instructed not to hold or defend the Copernican view. His defence was that the ban imposed in 1616 did not prevent teaching the Copernican view. He argued that his dialogue presented both views to the reader. The judges decided that Galileo had broken the 1616 ban and had acted deceitfully. He was threatened with torture, forced to recant and made to spend the rest of his life under house arrest at his home in Florence where he died in 1642. Science as a high-profile activity came to a standstill in Catholic Europe for many years. However, Galileo's dialogue and his discoveries were taken up vigorously by scientists in Northern Europe where the Church had much less authority. The publicity that Galileo's trial attracted perhaps helped to promote the Copernican system. Galileo's immense contribution to the development of science cannot be understated.

Motion in a straight line at constant speed

Speed is defined as distance travelled per unit of time. The scientific unit of speed is the metre per second (abbreviated as m/s or $m\,s^{-1}$).

For an object moving at constant speed, the distance travelled by the object in a certain time can be calculated by multiplying the speed by the time taken.

$$\text{Distance} = \text{speed} \times \text{time taken}$$

For example, suppose a car is travelling at a constant speed of 12 m/s. The car therefore travels a distance of 12 metres each second and a distance of 720 metres each minute (= 12 m/s × 60 seconds per minute). In 1 hour at this speed, the car would travel a distance of 43.2 kilometres (= 43 200 metres = 12 m/s × 3600 seconds per hour).

For an object with varying speed, its average speed over a certain distance or time is defined as the distance travelled / the time taken.

$$\text{Average speed} = \frac{\text{distance travelled}}{\text{time taken}}$$

Worked example

A train leaves a station at 11.00 a.m. and arrives at the next station 15 kilometres away at 11.22 a.m. Calculate (a) the time taken in seconds, (b) the average speed of the train in metres per second on this part of its journey.

Solution

(a) Time taken = 22 minutes = 22 × 60 s = 1320 s
(b) Average speed = 15000 m / 1320 s = 11.4 m/s

Speed limits

The speed limit on UK motorways is 70 miles per hour.

- 1 mile = 1.6 kilometres, so 70 miles = 70 × 1.6 = 112 kilometres. Thus a speed of 70 miles per hour = 112 kilometres per hour.
- 1 kilometre = 1000 metres and 1 hour = 3600 seconds, so a speed of 1 kilometre per hour = 1000 metres / 3600 seconds = 0.278 m/s. Thus a speed of 112 kilometres per hour = 31.1 m/s.
- Abbreviations: mph means 'miles per hour'; kph or km^{-1} means 'kilometres per hour'.
  Conversions: 5 mph = 8 km h^{-1} = 2.2 m/s.

Distance v. time graphs

The motion of an object can be represented by a graph of distance on the vertical scale (called the y-axis) against time on the horizontal scale (called the x-axis). For an object moving at constant speed in a straight line, the distance of the object from a fixed point changes by equal amounts in equal times. Hence the graph of distance v. time for such an object is a straight line as in Fig. 2A.

The steepness of the line depends on the speed of the object. In Fig. 2A, the steepness of the line is constant which means that the speed of the object is constant. The greater the speed of the object, the greater the distance the object moves each second so the steeper the line is. The steepness of the line is called its **gradient** and is defined as the change of the quantity plotted on the y-axis divided by the corresponding change of the quantity plotted on the x-axis.

Thus the gradient of the line = change of distance / time taken = the speed of the object.

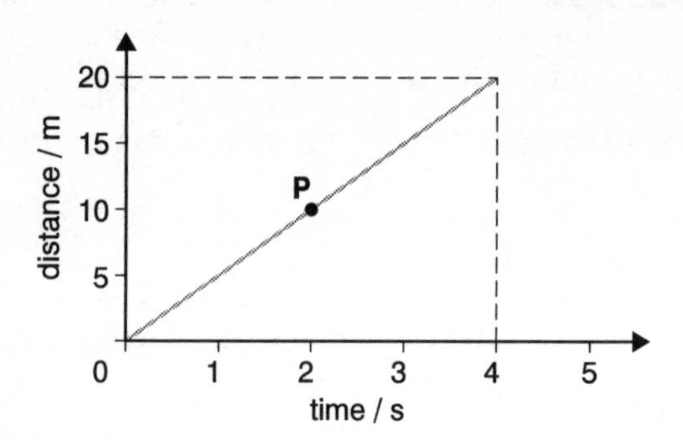

Figure 2A A distance v. time graph

Worked example

(a) Determine the speed of the object whose motion is represented by the line in Fig. 2A.
(b) How far would this object travel at this speed in (i) 1 minute, (ii) 30 minutes?

Solution

(a) The gradient of the line = 20 metres / 4 seconds = 5 m/s.
(b) (i) Distance moved in 1 minute = speed × time
= 5 m/s × 60 s = 300 m.
(ii) Distance moved in 30 minutes = speed × time
= 5 m/s × 30 × 60 s = 9000 m.

Equations and symbols

Because word equations such as ' distance = speed × time' become tedious and time-consuming, letters are used as 'shorthand' symbols for physical quantities so the above word equation is usually written as $s = \upsilon t$, where s represents distance moved, υ represents speed and t represents time taken. Each symbol stands for a certain physical quantity in accordance with an agreed scientific convention.

An equation may need to be rearranged by moving the symbols from one side to the other. The basic rule to remember is to do the same to both sides of the equation.

For example, in the equation $s = vt$, the symbol s is said to be the subject of the equation because it appears first as you read the equation from left to right. To make v the subject of the equation,

- divide both sides of $s = vt$ by t to give $\frac{s}{t} = \frac{vt}{t}$
- cancel t from the top and bottom of the right-hand side of the equation to give $\frac{s}{t} = v$
- swap the two sides of the equation over so v is read first; $v = \frac{s}{t}$

Velocity and speed

Velocity is defined as speed in a specified direction. The unit of velocity is the metre per second (m/s), the same as the unit of speed.

Two objects travelling at 60 mph in opposite directions along a motorway have the same speed but not the same velocity. An object moving along a circular path at constant speed has a continually changing direction of motion so its velocity continually changes even though its speed is constant.

Summary

- **Speed** is distance travelled per unit time
- **Velocity** is speed in a given direction
- For an object moving at constant speed, the distance moved in a certain time = speed × time
- 5 mph = 8 km h^{-1} = 2.2 m/s

Questions

Q1. Calculate the distance moved
(a) in 1 minute by a car travelling at a constant speed of 12 m/s,
(b) in 30 minutes by a plane travelling at a constant speed of 250 m/s,
(c) in 10 minutes by a car travelling at a constant speed of 30 m/s.

Q2. Calculate the time taken for
(a) a train moving at a constant speed of 15 m/s to travel a distance of 1 kilometre,

(b) a car travelling at a constant speed of 20 m/s to travel a distance of 500 m,
(c) a plane travelling at a constant speed of 150 m/s to travel a distance of 250 km.

Q3. Calculate the average speed of
(a) a train that takes 20 minutes to travel a distance of 15 km,
(b) an athlete who takes 10 s to run a distance of 100 m,
(c) a plane that takes 1 hour 40 minutes to travel a distance of 800 km.

Q4. Convert the following speeds into m/s:

(a) 30 mph (b) 700 mph (c) 100 km h^{-1} (d) 12 km min^{-1}

Q5. Two cars join a motorway at the same junction at the same time, travelling in the same direction.
One of the cars travels at 30 m/s for 40 minutes then stops at a service station for 20 minutes before rejoining the motorway just as the other car is passing the service station.
(a) Calculate the distance travelled by the first car from the junction where it joined the motorway to the service station,
(b) Show that the average speed of the other car was 20 m/s.

Accelerated motion along a straight line

Acceleration is defined as change of velocity per unit of time. The unit of acceleration is the metre per second per second, usually written as m/s^2 or m/s^{-2}.

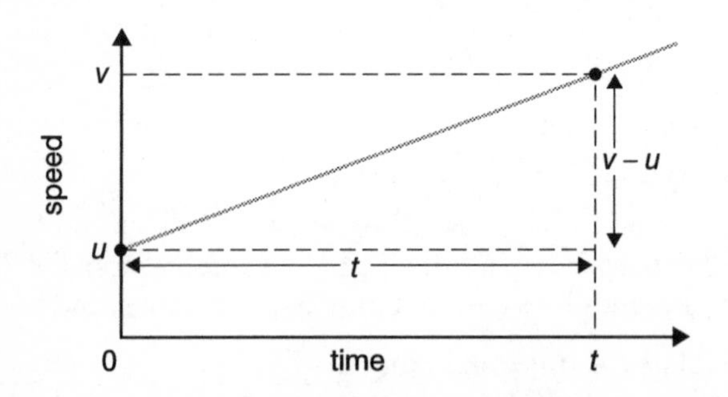

Figure 2B A speed v. time graph for constant acceleration

For an object moving in one direction along a straight line, its acceleration is constant if its speed changes by equal amounts in equal times. Suppose the speed of such an object changes from u to υ in time t. This change is shown by the straight line on the speed – time graph shown in Fig. 2B.

The change of speed in time $t = \upsilon - u$

$$\therefore \text{the acceleration, } a = \frac{\text{change of speed}}{\text{time taken}} = \frac{\upsilon - u}{t}$$

Rearranging this equation gives $a\,t = (\upsilon - u)$
hence $\upsilon = u + at$

Note: A negative value of acceleration is referred to as deceleration. This is the term used for slowing down.

Worked example

A car accelerates without change of direction from rest to a speed of 30 m/s in 60 s. Calculate its acceleration.

Solution

Initial speed $u = 0$, final speed $\upsilon = 30$ m/s, time taken $t = 60$ s

$$\therefore a = \frac{\upsilon - u}{t} = \frac{30 - 0}{60} = 0.5 \text{ m/s}^2$$

More about speed v. time graphs

1 The acceleration of an object = the gradient of the line.
The gradient of the line is the change of the quantity plotted on the y-axis (i.e. speed) divided by the change of the quantity plotted on the x-axis (i.e. time taken). In Fig. 2B, the gradient = $(\upsilon - u) / t$ which is the acceleration of the object.

2 The distance travelled = the area between the line and the x-axis. In Fig. 2B, the average speed of the object over time $t = \frac{1}{2}(u + \upsilon)$ because the speed increases steadily from u to υ.

3 The distance travelled in time t, s = the average speed × the time taken

hence $s = \frac{1}{2}(u + \upsilon)\,t$

The area between the line and the x-axis in Fig. 2B is a trapezium with a base representing t and with sides of heights represented by u and υ. This shape has the same area as a rectangle of height corresponding to $\frac{1}{2}(u + \upsilon)$ and base t, thus the area corresponds to the distance travelled $\frac{1}{2}(u + \upsilon)\,t$.

Worked example

A train on a straight line accelerates from rest to a speed of 20 m/s in a time of 100 s then travels at constant speed for 250 s. It then decelerates at a constant rate for 50 s and stops.

(a) Draw a speed v. time graph to represent the motion of the train.
(b) Calculate the acceleration and the distance travelled in (i) the first 100 s, (ii) the next 250 s, (iii) the final 50 s.
(c) Calculate the average speed of the train over the journey.

Solution

(a) See Fig. 2C

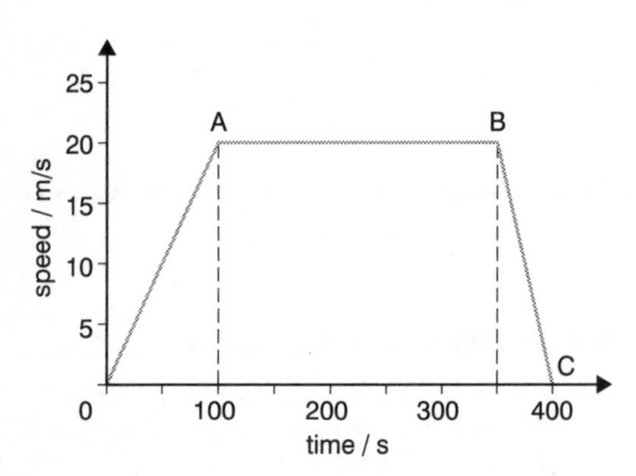

Figure 2C

(b) (i) 1st stage (0 to 100 s); $u = 0$, $\upsilon = 20$ m/s, $t = 100$ s

$$\therefore a = \frac{\upsilon - u}{t} = \frac{20 \text{ m/s} - 0}{100 \text{ s}} = 0.2 \text{ m/s}^2$$

$s = \frac{1}{2}(u + \upsilon)\,t = \frac{1}{2}(0 + 20 \text{ m/s})\ 100 \text{ s} = 1000$ m

(ii) 2nd stage (100 s to 250 s); $u = \upsilon = 20$ m/s, $t = 250$ s

$a = 0$ because there is no change of its speed,
$s = 20 \text{ m/s} \times 250 \text{ s} = 5000$ m

(iii) 3rd stage (250 to 300 s); $u = 20$ m/s, $\upsilon = 0$, $t = 50$ s

$\therefore \quad a = \frac{\upsilon - u}{t} = \frac{0 - 20 \text{ m/s}}{50 \text{ s}} = -0.4 \text{ m/s}^2$

$s = \frac{1}{2}(u + \upsilon)\, t = \frac{1}{2}(20 + 0) \times 50 = 500 \text{ m}$

(c) Total distance travelled = 1000 + 5000 + 500 = 6500 m

Total time taken = 100 + 250 + 50 s = 400 s

$\therefore$ average speed = $\frac{\text{total distance}}{\text{time taken}} = \frac{6500 \text{ m}}{400 \text{ s}} = 16(.25)$ m/s

Galileo and gravity

Galileo used an inclined plane to show that an object rolling down an inclined plane gathered speed at a constant rate as it descended. He showed that the acceleration was constant and could be made larger by increasing the steepness of the incline.

He thought that the acceleration of an object falling freely ought to be constant and ought to be the same for any object. However, he was unable to time the vertical descent of an object because his clock was not accurate enough.

One method of timing the descent of a falling object is to photograph a small steel ball as it falls in front of a vertical scale. If the apparatus is in a darkened room that is illuminated by a flashing stroboscope, an image of the ball is captured on the photograph each time the light flashes. Provided the light flashes at a constant rate, the photograph therefore shows the position of the ball at equal intervals. A video camera could be used instead of a film camera but the stroboscope would still be needed. If the flash rate of the stroboscope is known, the time interval between successive flashes can be worked out. Figure 2D shows the idea. Measurements from such an investigation are shown below.

Time from start / s	0.0	0.1	0.2	0.3	0.4	0.5
Distance fallen / m	0.0	0.05	0.20	0.45	0.80	1.25
Average speed / m s^{-1}	0.0	0.50	1.00	1.50	2.0	2.5
Speed / m s^{-1}	0.0	1.00	2.00	3.00	4.00	5.00

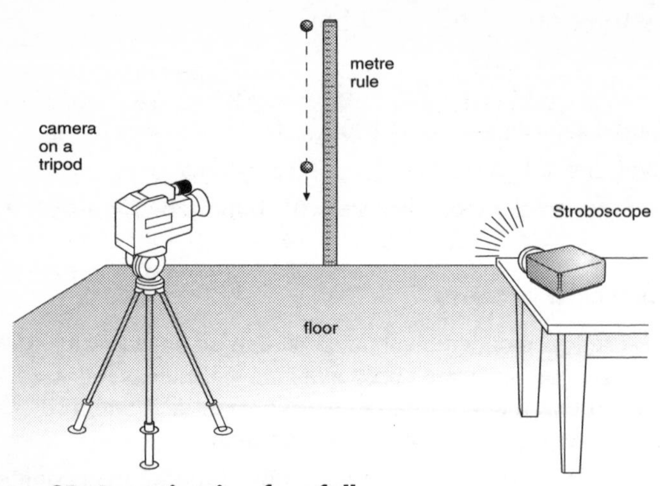

Figure 2D Investigating free fall

The average speed between the start and each flash has been worked out in the last column by dividing the distance fallen by the time taken from the start. Because the initial speed is zero, the speed at each flash is twice the average speed. The results show that the speed increases by 1.0 m/s every 0.1 s which scales up to an increase of 10 m/s every second. Thus the acceleration of free fall has a constant value of 10 m/s^2. This value is usually referred to as g, the acceleration due to gravity of a freely falling object. Accurate measurement of g shows that g varies slightly with latitude. Its value is 9.81 m/s^2 at the Earth's poles and 9.78 m/s^2 at the Equator.

Summary

- Acceleration is change of velocity per unit time.
- The acceleration due to gravity, $g = 9.8$ m/s^2 at the Earth's surface.
- For an object which undergoes constant acceleration a in a straight line from initial speed u to speed υ in time t covering a distance s,

$$\upsilon = u + at$$
$$s = \tfrac{1}{2}(u + \upsilon)\,t.$$

Questions

Q6. A car accelerates from rest at an acceleration of 0.5 m/s^2 for 20 seconds. Calculate (a) its speed after 20 seconds from rest, (b) the distance it moves in this time.

Q7. In an experiment, a ball was released from rest at the top of a slope and rolled a distance of 2.0 metres down the slope in 8.0 seconds. Calculate (a) its average speed in this time, (b) the speed after 8.0 seconds from rest, (c) the acceleration of the ball.

Q8. An aircraft accelerates from rest to a speed of 120 m/s in 40 s when it takes off. Calculate
(a) its average speed during take off,
(b) the distance it travels in this time,
(c) its acceleration during this time.

Q9. A small object is released from the top of a well and hits the water surface of the well 1.5 s later. Calculate (a) its speed just before it hits the water surface, (b) the distance from the top of the well to the water surface. Assume $g = 9.8\ m/s^2$.

Q10. A rocket launched vertically accelerated at a constant rate for 50 s from the moment it was launched to when its fuel ran out when its speed had reached 200 m/s.
(a) Calculate the acceleration of the rocket during this time.
(b) Calculate the height of the rocket when its fuel ran out.
(c) Explain why the rocket continued to gain height for a further 20 s (to the nearest second) after its fuel ran out.
(d) Calculate the maximum height attained by the rocket. Assume $g = 9.8\ m/s^2$.

3 FORCES IN ACTION

In the two centuries after Galileo's death, science developed at an ever-increasing rate. By the end of the nineteenth century, many scientists believed that all the fundamental principles of physics had been discovered and little else remained except to make more accurate measurements of the properties of matter. A few niggling issues remained that could not be properly explained, mostly in connection with light, but it was generally thought that the problems would be solved using the known principles of physics. The laws of mechanics had been established by Sir Isaac Newton in the late seventeenth century and had proved enormously successful in explaining the motion of objects from the throw of a ball to the motion of the planets in orbit round the Sun. Michael Faraday had discovered how to generate electricity in the 1820s and, by the end of the century, his ideas led to the establishment of the electricity supply industry.

The foundations of physics that seemed so secure and certain at the end of the nineteenth century fell apart in the first two decades of the twentieth century. We will look at how and why this happened later in this book. Einstein's theories of relativity and the quantum theory reshaped the principles of the subject and new discoveries about the properties of matter have resulted in major new industries such as electronics, nuclear power and aerospace. However, pre-twentieth century physics which is usually referred to as classical physics arguably had an even greater impact because the principles of classical physics underpinned the great developments in engineering and construction that created the industrial revolution of the nineteenth century. In this chapter, we will consider in detail the laws of motion established by Newton and how these classical laws are used in some everyday situations.

Newton's laws of motion

Galileo's ideas on motion and the Copernican system were brought together by Sir Isaac Newton in his laws of motion and gravitation. For over two centuries, these laws were thought to be universal, applicable in any situation where material bodies interact. Newton's three laws of motion underpin much of modern technology and engineering. The motion of planets, comets and satellites can be explained almost completely using Newton's theory of gravitation and his laws of motion.

Before Newton wrote his laws of motion, he defined key physical quantities which he then linked in his laws of motion to each other and to other physical quantities already defined.

1 The quantity of matter in an object or its **mass** was defined by Newton as its volume × the density of the substance from which the object is composed. For example, iron is about eight times as dense as water so the mass of 2 cubic metres of steel is sixteen times the mass of 1 cubic metre of water. The mass of an object is now determined by comparing its weight with the weight of a standard mass. See p. 15.

2 The quantity of motion of a moving object or its **momentum** was defined by Newton as its mass × its velocity. For example, the momentum of an object of a 2 kg mass moving at 3 m/s is six times the momentum of an object of mass 1 kg moving at 1 m/s.

3 A force is an action on an object that changes its motion. Two or more forces acting on an object balance each other out if the object stays at rest or moves at constant speed along a straight line.

Newton's 1st Law of motion

An object continues at rest or in uniform motion unless acted on by a force.

Here are some everyday situations that reveal the link between force and motion as established by Newton in his 1st Law:

1 An object sliding across ice moves at constant speed without change of direction because there is no force on the object due to the ice. If the object crosses a surface where there is no ice, it is brought to a standstill by the force of friction between the object and the surface.

2 A cyclist in motion on a level road gradually comes to a halt if he or she stops cycling and 'freewheels'. The reason is that friction in the wheel bearings and air resistance on the cyclist act against the motion and gradually stop the cycle moving.

3 An object being whirled round on the end of a length of string is pulled by the string so its direction repeatedly changes. If the string suddenly snaps, the object flies off at a tangent because the pull force of the string on the object has suddenly stopped acting on the object.

In effect, Newton's 1st Law tells us what a force is, namely anything that can change the motion of an object. Different types of forces include:

- the force of gravity on an object, usually referred to as the weight of the object,
- the force of friction acting on two surfaces when they slide over each other,
- the tension in an object such as a string when it is taut and being pulled at each end,

force F_1 force F_2

In a tug-of-war 'stalemate', the teams pull with equal and opposite forces. The forces balance each other out. $F_1 = F_2$

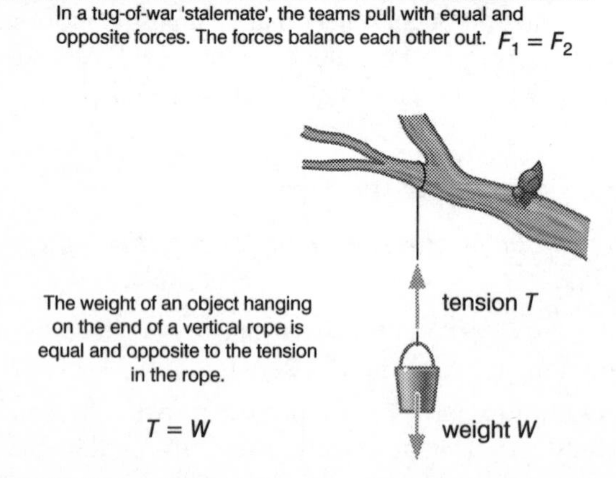

The weight of an object hanging on the end of a vertical rope is equal and opposite to the tension in the rope.

$T = W$

Figure 3A Balanced forces

- the force on an object when another object pushes or pulls it or supports it,
- the force between two electrically charged objects or between two magnetized objects.

Balanced forces

An object acted on by two or more forces can be at rest or in uniform motion if the forces on it balance each other out. The simplest example of this situation is a tug-of-war when a stalemate exists because the two teams pull on the rope with equal force in opposite directions. Another example is when a rope hanging vertically supports a bucket. The force of gravity on the bucket acts downwards and is equal and opposite to the tension in the rope which acts upwards, as shown in Fig. 3A.

Newton's 2nd Law of motion

The force on an object is proportional to the rate of change of momentum of the object.

An object initially at rest acted on by a single force gains speed and therefore gains momentum. For a given amount of force, the object gains a certain amount of momentum every second during the time the force acts on it. Consider the example of an object of mass m being accelerated along a straight line by a constant force F for a time t acting in the same direction as the object is moving. Suppose the object's speed increases from u to υ in this time without change of direction.

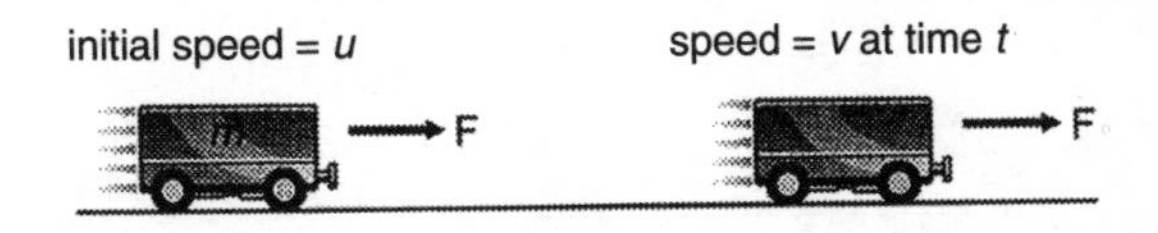

Force and motion

Figure 3B Newton's 2nd Law of motion

- The initial momentum of the object = mu
- The final momentum of the object = $m\upsilon$

so the change of momentum = final momentum – initial momentum = $m\upsilon - mu$

hence the rate of change of momentum =

$$\frac{\text{change of momentum}}{\text{time taken}} = \frac{m\upsilon - mu}{t} = \frac{m(\upsilon - u)}{t}.$$

According to Newton's second law, the force F on the object is proportional to the rate of change of momentum, hence $F = k\,m\,\frac{(\upsilon - u)}{t}$ where k is a constant of proportionality.

Because the acceleration of the object $a = \frac{(\upsilon - u)}{t}$, then $F = k\,m\,a$.

By defining the unit of force, the **newton**, as the amount of force that would give an object of mass 1 kg an acceleration of 1 m/s^2, the value of k is therefore set at 1. Newton's 2nd Law can then be written as the equation

$F = m\,a$ where F = force in newtons,
m = mass in kilograms, and
a = acceleration in m/s^2

Worked example

A car of mass 600 kg accelerates from rest to a speed of 12 m/s in 30 s. Calculate (a) its acceleration, (b) the force needed to produce this acceleration.

Solution

(a) Initial speed, $u = 0$, final speed $\upsilon = 12$ m/s, time taken, $t = 30$ s,

$\therefore$ acceleration $a = \frac{(\upsilon - u)}{t} = \frac{12 - 0}{30} = 0.4\ m/s^2$,

(b) $F = ma = 600\ kg \times 0.4\ m/s^2 = 240$ N.

Activity

Here are some simple experiments to demonstrate Newton's first two laws of motion.

1 *Newton's 1st Law of motion*: Flick a coin across a very smooth level surface (e.g. a tea tray) and it will slide across the surface.

In the absence of friction, the coin continues to move without being pushed because no force acts on it.

2 *Newton's 2nd Law of motion*: Release a coin at the top of a flat slope (e.g. a tea tray propped up at one end) so that it slides down the slope. Observe it gather speed as it moves down the slope. Its momentum increases because its weight provides a steady force that pulls it down the slope.

Sir Isaac Newton 1642–1727

The ideas about machine power were first developed by Sir Isaac Newton when he applied his ideas about force and motion to machines. Newton was born in the market town of Grantham in the county of Lincolnshire in eastern England. His father died before he was born and he was brought up by a grandparent after his mother remarried. He was sent to the local grammar school as a boarder and entered Cambridge University in 1661. England at this time was a republic under Oliver Cromwell. The University was closed at times during 1665 and 1666 because of the Great Plague which ravaged the country during those years. Newton returned to his home in the Lincolnshire countryside and in just two years produced mathematical theorems and physical theories that revolutionized mathematics and physics. He returned to Cambridge in 1667 and was appointed two years later at 26 years of age to the Chair of Mathematics at Trinity College.

He published his theories of mathematics and physics in 1687 in his greatest work, the Principia, in which he showed that his three laws of motion and his law of gravitation are sufficient to explain the motion of any system of bodies. He proved once and for all that the planets and the Earth orbit the Sun and he explained Kepler's laws of planetary motion and Galileo's observations on the motion of falling objects. Using his law of gravitation which we will meet in Chapter 13, he was able to predict comets, eclipses and tides. He saw the Universe as a gigantic mechanical system, governed by the same laws that govern the motion of objects on the Earth. His ideas provided the guiding principles for science for the next two centuries until Einstein showed that space and time are not independent quantities. We will meet Einstein's ideas in detail in Chapter 10.

Newton's interests in science were wide-ranging, and included astronomy, chemistry and optics as well as mathematics and physics. He left the University in 1696 to become Master of the Mint where he devoted his talents to monetary reform, only delving back to the challenges of science occasionally. His pre-eminence as a scientist was recognized in 1703 when he was elected as President of the Royal Society and he was knighted in 1705. In contrast with Galileo who was shunned by the Catholic Church, Newton became part of the Establishment in England, even attracting the attention of satirists, clearly unaware that the freedom of thought on which they depended would not have emerged without Galileo and Newton.

Mass and weight

The acceleration of a freely falling object, g, is constant, provided the distance fallen is much smaller than the radius of the Earth. This acceleration is caused by the **weight** of the object which is the force of gravity pulling it towards the Earth. Using Newton's 2nd Law in the form $F = m\,a$ therefore gives $m\,g$ as the force of gravity on an object of mass m. Hence the weight W of an object of mass $m = mg$.

weight W (in newtons) $= m\,g$, where m = mass in kilograms and g = the acceleration of a freely falling object (= 9.8 m/s^2 near the Earth's surface).

The weight of a 1 kg mass near the Earth's surface is therefore 9.8 N. The weight of a person of mass 60 kg at the Earth's surface is 588 N.

The gravitational field strength at any position is defined as the force of gravity per unit mass on an object at that position. Hence gravitational field strength $= mg / m = g$. Thus g can be referred to as either the acceleration of a freely falling object (in m/s^2) or gravitational field strength (in newtons per kilogram).

Newton's 3rd Law

When two bodies interact, they exert equal and opposite forces on each other.

For example, if you clap your hands together, the force of your left hand on your right hand is equal and opposite to the force of your right hand on your left hand.

If you lean on a wall, the wall exerts a force on you equal and opposite to the force you exert on the wall. When you stand still, the floor exerts an upward force on you equal and opposite to the force you exert on the floor.

Summary

Momentum = mass × velocity

Newton's laws of motion

Newton's 1st Law: An object continues at rest or in uniform motion unless acted on by a force.

Newton's 2nd Law: Force = mass × acceleration.

Newton's 3rd Law: When two bodies interact, they exert equal and opposite forces on each other.

Weight = mass × g

Questions $g = 9.8\ \text{m/s}^2$

Q1. Explain why a car is likely to skid at a bend on an icy road.

Q2. (a) Calculate the weight of an object of mass (i) 2 kg, (ii) 4 kg.
(b) Explain why a 2 kg object and a 4 kg object released above the ground at the same time fall at the same rate.

Q3. (a) A bucket of mass 2.5 kg contains 0.008 m^3 of water.
(i) Calculate the mass of water in the bucket. The density of water is 1000 kg m^{-3}.
(ii) Calculate the total weight of the bucket and the water.
(b) The bucket is attached to a rope which is used to raise the bucket. Calculate the tension in the rope when it supports the bucket at rest containing 0.008 m^3 of water.

Q4. Calculate
(a) the force needed to give an object of mass 4 kg an acceleration of (i) 2 m/s^2, (ii) 6 m/s^2.
(b) the acceleration of an object of mass 2 kg acted on by a force of (i) 10 N, (ii) 8 N.

Q5. A car of mass 800 kg is brought to rest from a speed of 30 m/s in a time of 15 s. Calculate
(a) the deceleration of the car,
(b) the force needed to produce this deceleration.

Stability

Force as a vector

A physical quantity that has a direction as well as a size is a **vector** quantity. Examples of vectors include force, velocity, acceleration and momentum. A physical quantity that is not directional is referred to as a **scalar** quantity. Examples include distance, speed, mass and energy.

A vector quantity may be represented by an arrow of length in proportion to the quantity and which points in the appropriate direction. For example, a force of 10 N acting horizontally due North on an object may be represented on a diagram as an arrow of length 10 cm pointing in a direction defined on the diagram as due North. Caution is needed to use Newton's 3rd Law correctly when force diagrams are drawn. For every force acting on a body, there is an equal and opposite force exerted by the body. To ensure force diagrams do not become complicated, a force diagram for an object should show only the forces acting on the body not the forces exerted by the body.

Centre of gravity

When an object is acted upon by two or more forces including the force of gravity, the effect on the object depends on the direction and size of the forces. Each force acting on an object may be represented as a vector on a force diagram. To represent the force of gravity on an object (i.e. its weight) which acts on all parts of an object, we define the centre of gravity of an object as the point where its entire weight may be considered to act. The weight of an object can then be shown on a force diagram as a vector arrow acting downwards at the centre of gravity.

The idea of centre of gravity enables us to understand stability and the conditions necessary to keep an object in equilibrium. The examples below are chosen to illustrate the idea:

1 A tall free-standing bookcase is liable to topple over when it is moved unless it is supported. If it tilts too much, its centre of gravity moves outside its base and it topples over unless someone supports it to prevent it falling over. Push a tall object such as an empty cereal packet sideways at its top and you will find it tilts then topples over if it is pushed too far. If you hold it at the point where it only just topples, you should find the centre of the box is directly above the edge in contact with the table.

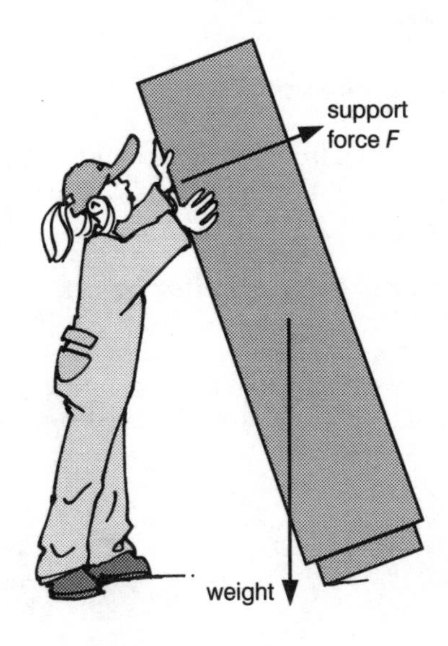

Figure 3C Toppling

2 A beam supported on a pivot which is not at the centre of the beam can be balanced by hanging a single weight from the beam between the pivot and the end of the beam nearest the pivot. Figure 3D shows the arrangement.

The centre of gravity of a uniform beam is at its centre. Using the principle of the lever (see p.13)

$$W_1\, d_1 = W_0\, d_0$$

where W_1 is the weight of the hanging object, W_0 is the weight of the beam, d_1 is the distance from W_1 to the pivot and d_0 is the distance from the pivot to the centre of gravity of the beam.

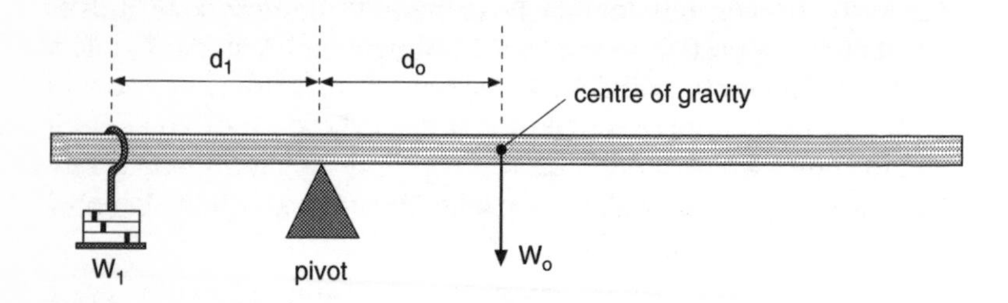

Figure 3D Balancing a beam off-centre

Note that d_0 is the distance from the pivot to the centre of gravity of the beam. Hence the weight of the beam, W_0, can be calculated if W_1, d_0 and d_1 are known.

Locate the centre of gravity of a flat object

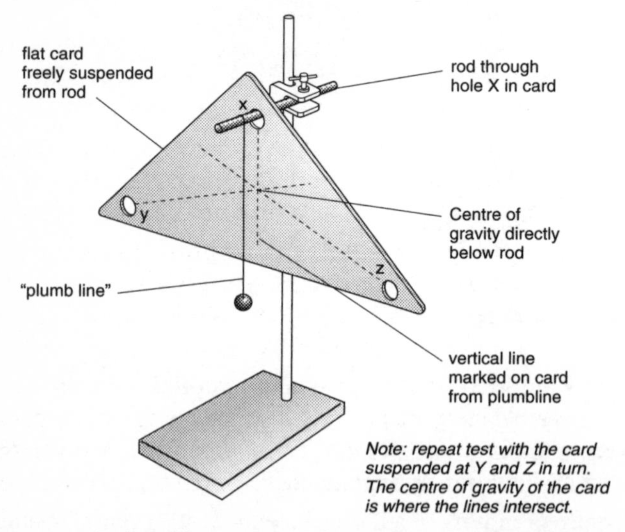

Figure 3E Finding the centre of gravity of a flat object

Cut a piece of card to make a large triangle. Pierce a small hole near one corner of the triangle and hang it freely from a pin. Hang a 'plumbline' consisting of a cotton thread supporting a small weight from the pin against the triangle. Draw a pencil line on the triangle along the line of the thread. When the triangle is at rest, its centre of gravity is directly below the pin along this line. Repeat the test by hanging the triangle from a pinhole at each of the other two corners in turn. The centre of gravity of the card triangle is where the lines through the pinholes intersect.

Summary

A **vector** quantity has magnitude (i.e. size) and direction. A **scalar** quantiy has magnitude only.

The centre of gravity of an object is the point where its entire weight may be considered to act.

Questions

Q6. Sketch a free body force diagram for
(a) a bucket at rest hanging on the end of a vertical rope,
(b) a tightrope walker balanced on a tightrope.

Q7. (a) Explain why a free-standing bookcase topples over easily if only the top shelf is filled with books.
(b) Explain why a person carrying a heavy back pack is liable to topple backwards when sitting on a wall unless the backpack is removed.

Q8. A spanner is a device designed to tighten or untighten a nut on a bolt. Explain why less effort is needed to untighten a nut using a long spanner than is needed using a short spanner.

Q9. A child's seesaw consists of a uniform beam of length 4.0 metres pivoted at its centre. The seesaw is balanced when a girl sits at one end and a boy of weight 200 N sits on the seesaw 1.0 m from the other end.
(a) Sketch the force diagram for the seesaw,
(b) Calculate the weight of the girl.

Q10. A uniform beam of length 5.0 m is positioned at right angles to the edge of a river bank so that a 2.0 m section of the beam projects from the bank.
(a) Explain why the beam does not fall into the river in this position.
(b) Why would the beam topple into the river if someone attempted to walk on it beyond a certain distance from the bank?

4 MACHINES AT WORK

The Industrial Revolution in the eighteenth and nineteenth centuries developed because engineers were able to make bigger and more powerful machines and engines than the windmills and watermills of the pre-industrial era. The scientific principles established by Newton enabled engineers to design and build such machines and engines. With Newton's guiding principles, engineers continue to seek more efficient and cost-effective solutions to modern problems. In this chapter we look at how the ideas established by Newton are used to explain how different types of machines work.

Work, energy and power

When a lever is used to shift an object, for example to prise open a lid, a force applied by hand at the end of the lever has a much larger effect via the lever than if it was applied directly. The greater the distance from the pivot to the point where the force is applied, the greater the effect. A lever is a force multiplier. The effect of the applied force is multiplied so a much larger force acts. Another example of a lever at work is a manual punch used to punch a hole in sheets of paper. When a downward force is applied by hand to the lever, the metal cutter is pushed through the paper. The force of the hand on the lever is multiplied tenfold because this force acts ten times further from the pivot than the metal cutter.

Although most devices used to move objects are more complicated than a simple lever, the same general principle applies. The device exerts a force to move an object (which we will refer to as the load) as a result of a force (which we shall refer to as the effort) applied to the machine or device. The effort needed to shift a given load is

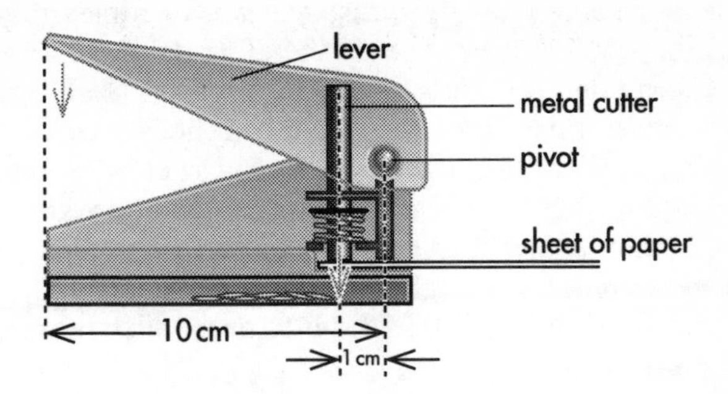

Figure 4A A manual punch

smaller than the load because of the way the device is designed. For example, a heavy weight supported by a rope attached to the drum of a winch can be raised by turning the handle of the winch. When the handle is turned, the rope winds round the drum thus raising the weight attached to the rope. The effort applied to the handle acts at a much larger distance than the distance from the rope on the drum to the centre of the drum. The pull of the rope on the drum is the load that has to be shifted by the effort. Figure 4B shows an end-view of the drum. The law of levers (see p. 14) can be applied here to give the equation

$$F_E d_1 = W d_2,$$

where F_E is the effort, d_1 is the distance from the line of action of the effort to the axis of the drum, W is the weight to be raised and d_2 is the radius of the drum. Since d_1 is much greater than d_2, then F_E is much smaller than W.

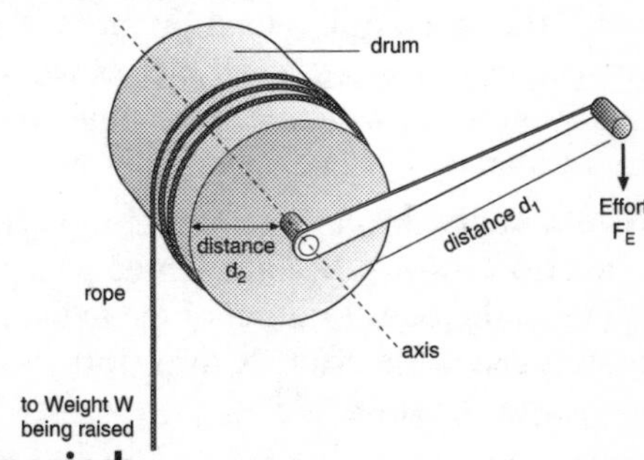

Figure 4B Using a winch

Suppose in the above example, distance d_1 is ten times greater than distance d_2. A 500 N weight could therefore be raised by an effort of 50 N, provided no other forces are present. Each turn of the handle would raise the weight by a height equal to the circumference of the drum which is $2\pi d_2$. The effort is applied at a much greater distance d_1 from the axis, so the effort goes round a circle of circumference $2\pi d_1$ which is much further than the height gain of the weight for a single turn. Thus the reduction of effort is at the expense of an increase of distance over which the effort must be applied.

In a machine where there is no friction, the ratio of the distance moved by the effort to the distance moved by the load is equal to the ratio of the load to the effort. In other words, if the load is 10 times the effort, the effort must move a distance of 10 metres for every metre the load is moved through. If the machine is operating at a steady speed, the effort must therefore move round at ten times the speed at which the weight rises. For example, suppose the handle is turned so it goes round at a steady speed of 1 m/s, the weight would therefore rise at a steady speed of 0.1 m/s for a load / effort ratio of 10. Thus for a given load,

the load × the load speed = the effort × the effort speed.

The **power** of a machine depends on the load it moves and the speed at which it moves the load. A machine that shifts four times as much load as another machine but does the job twice as fast as the other machine must be eight times more powerful than the other machine.

- The power of a machine in operation, in *watts*, is the load (in newtons) × the speed of the load (in m/s). A machine that shifts a load of 100 N at a speed of 10 metres per second has a power output of 1000 watts which is 1000 times more than a machine that shifts a load of 1 N at a speed of 1 m/s.
- The power supplied to a machine is the effort × the speed at which the effort moves. For a machine which is 100% efficient, this is equal to the power output of the machine because the load × the load speed = the effort × the effort speed. In other words, provided no power is wasted

  the power output of a machine = the power supplied to it

Work and energy

The **work** done by a force on an object is defined as the force × the distance moved by the object along the line of action of the force. The unit of work is the *joule* (J) which is equal to the work done when a force of one newton moves its point of application through a distance of one metre in the direction of the force.

Work done = force × distance moved in the direction of the force
(in joules) (in newtons) (in metres)

For example, the work done on a brick of weight 2.0 N lifted through a height of 1 m is equal to 2 J. If the same brick had been raised through only 0.5 m, the work done on it would have been 1 J. The work done on an object by raising it enables the object to do work on something else when it drops. A brick dropped onto someone's shoe would do work in denting the shoe. The work done by the brick on the shoe could not be recovered, however, unless the shoe cap was elastic. A raised object has the capacity to do work because of its position. Another example is a clockwork spring which has work done on it when it is wound up. When the spring unwinds, it does work by making the wheels and gears inside the clock move. If the spring is prevented from unwinding, it keeps the capacity to do work until it is allowed to unwind. This capacity to do work is defined as the **energy** of an object. The term 'stored work' might seem more appropriate but work is only done when a force moves so the term 'energy' is used for the capacity of a body to do work.

The energy of an object is its capacity to do work.

Energy is measured in joules, the same unit as the unit of work. For example, if a brick of weight 5 N is raised by 2 m from the ground, the work done to raise it is equal to 10 J (= 5 N × 2 m). The raised brick therefore contains an extra 10 J of energy due to being lifted. If the brick is dropped onto someone's shoe, it causes a dent in the shoe and loses the extra 10 J of energy in the process. Thus 10 J of work must have been done to create the dent. You can work out the force of the impact from the depth of the dent by recalling that work done = force × distance. For example, a dent of 0.5 cm (= 0.005 m) would correspond to a force of 2000 N (= 10 J / 0.005 m) and a really big ouch!

Forms of energy

Objects can possess energy in various ways, referred to as 'forms' of energy. For example:

- kinetic energy is the energy of a moving object due to its motion. The faster an object moves, the more kinetic energy it has. An unbalanced force is needed to make an object move faster so the work done by the force is equal to the gain of kinetic energy of the object.
- potential energy is the energy of an object due to its position. For example, an object raised from the ground gains potential energy because a lift force opposite to the weight of the object is needed to raise the object.
- elastic energy is the energy stored in an object when it is stretched or squeezed. When the object returns to its undistorted shape, the elastic energy is released.
- electrical energy is the energy of particles such as electrons that carry electric charge. A capacitor is a device used to store electric charge. Energy is stored in this process because work must be done to force electrons into a capacitor because identical charged particles repel each other.
- thermal energy is the energy of an object due to its temperature. For example, in a car engine, the petrol-air mixture in a cylinder is ignited by means of a spark which causes the air to become very hot very quickly. The pressure of the hot air shoots up and forces the piston out so it turns the drive shaft. See p. 77.
- nuclear energy is energy that can be released from an atom which has an unstable nucleus. See p. 177.
- chemical energy is the energy that can be released when chemicals react. For example, when a battery is used to light a torchbulb, chemical energy is released inside the battery to force electricity through the torchbulb.

Other forms of energy include sound energy and light energy. Heat transfer is energy transfer due to temperature difference. When a hot object is in contact with a cold object, heat transfer from the hot object to the cold object reduces the thermal energy of the hot object and increases the thermal energy of the cold object. See p. 64.

The Principle of Conservation of Energy

When energy is transferred between objects in an isolated system, the total energy of all the objects after the change is the same as the total energy of all the objects before the change. The total energy of all the objects is unchanged (i.e. conserved).This is known as the Principle of Conservation of Energy. Energy is the capacity to do work. The principle of conservation of energy means that the total capacity of an isolated system of objects is unchanged by interactions between the objects. The principle has been tested many times and the measurements always come up with the result that the total energy is unchanged. When energy transfer within the system takes place, individual objects do work on each other so gaining or losing energy. However, the total capacity to do work is unchanged. In other words, the total energy is unchanged.

Some thoughtful experiments

- A pendulum bob on a string swings from side to side repeatedly. The pendulum bob loses potential energy and gains kinetic energy when it moves towards the centre. When it moves away from the centre, it loses kinetic energy and gains potential energy. Its total energy stays the same as its kinetic energy plus its potential energy at any point always add up to the same.
- When an object crashes to the floor without rebounding, it loses all its kinetic energy. What happens to this energy? A test using a mass of lead pellets dropped down a long tube shows that the temperature of the pellets increases. The kinetic energy of the pellets becomes thermal energy.
- When a small weight is used to raise a large weight at a steady speed using pulleys, the large weight gains potential energy and the small weight loses potential energy. If friction is negligible, the gain of potential energy of the large weight is equal to the loss of potential energy of the small weight. If friction is not negligible, the large weight gains less potential energy than the small weight loses. In other words, there is a discrepancy between the loss of potential energy of the small weight and the gain of potential energy of the large weight. This discrepancy was investigated by James Joule in the nineteenth century. He made careful measurements and showed that the discrepancy is

exactly accounted for by thermal energy. The unit of energy and work is named after Joule because his investigations and others that followed led to the conclusion that energy is conserved in the changes investigated. All the available evidence indicated that energy is always conserved. The Principle of Conservation of Energy was thus established and remains so to this day although future discoveries might disprove the principle.

Power and energy

On p. 50, we saw that the power output of a machine is given by the load × the speed at which the machine moves the load. This link can be applied to any object or machine being driven at constant speed. If the force is in newtons and the speed in metres per second, the power is in watts. For example, suppose the engine of a vehicle moving at a constant speed of 25 m/s exerts a driving force of 400 N. The power output of the engine is therefore 400 N × 25 m/s = 10 000 watts. In energy terms, the vehicle moves a distance of 25 metres every second so the work done by the engine each second is therefore 10 000 joules (= 400 N x 25 m). The power output in watts is therefore the work done by the engine per second. In other words, one watt of power is an energy transfer rate of one joule per second. This general statement applies to *any* energy transfer process.

Power is rate of transfer of energy.

The unit of power is the watt, (W), equal to a rate of transfer of energy of 1 joule per second. A 3000 watt electric heater emits heat at a rate of 3000 joules every second. A 100 watt light bulb emits light energy at a rate of 100 joules per second. Every minute, a 100 W light bulb would emit 6000 joules of light energy. A weightlifter who raises a 800 N weight through a height of 1 metre in 2 seconds has a power output of 400 watts because the weight is supplied with 800 J of potential energy (= weight × height gain) in 2 seconds.

$$\underset{\textit{(in watts)}}{\textit{Power}} = \frac{\textit{energy transferred (in joules)}}{\textit{time taken (in seconds)}}$$

The above equation may be rearranged to give

energy transferred (in joules) = power (in watts) × time taken (in seconds).

How powerful are you?

Time how long it takes you to walk up a flight of stairs that you normally use. If you cannot use stairs, time a friend walking who can use the stairs. Then measure the height from the bottom to the top stair. If necessary, measure the height of a single step and multiply by the number of steps.

To calculate your power output, multiply your weight in newtons by the total height gain to give your gain of potential energy. Then divide the potential energy gain by the time taken to give your power output in watts. Note that your weight in newtons is your mass in kilograms $\times$ g. For the purpose of this activity, use $g = 10 \text{ m/s}^2$.

Worked example

A person of mass 54 kg walks up a flight of 20 stairs in 6 seconds. The height of each stair is 0.15 m. Calculate the power output of this person.

Solution

Total height gain = $20 \times 0.15 \text{ m} = 3.0 \text{ m}$

Weight = $54 \text{ kg} \times 10 \text{ m/s}^2 = 540 \text{ N}$

Gain of potential energy = $540 \text{ N} \times 3.0 \text{ m} = 1620 \text{ J}$

Power output = $\dfrac{1620 \text{ J}}{6\ s} = 270 \text{ W}$

Summary

Work done (in joules) = Force (in newtons) $\times$ distance moved in the direction of the force (in metres)

Energy is the capacity to do work.

Power (in watts) = $\dfrac{\text{energy transferred (in joules)}}{\text{time taken (in seconds)}}$

Questions $g = 9.8 \text{ m/s}^2$

Q1. Describe the energy changes that take place when

(a) a pendulum bob swings from one side to the other side,

(b) a ball is released above a hard floor, hits the floor and rebounds to a lesser height.

Q2. (a) Calculate the gain of potential energy when a 500 N weight is raised by a height of 4.0 m.
(b) How much light energy is released by a 100 watt light bulb in 20 seconds?

Q3. (a) Calculate the electrical energy used by a 3000 watt electric kettle in exactly 5 minutes.
(b) How long would a 100 watt light bulb need to be switched on to release the amount of energy calculated in (a)?

Q4. A child of mass 40 kg climbs 3.0 m up a vertical rope in 25 seconds. Calculate (a) the weight of the child,
(b) the gain of potential energy of the child,
(c) the power output of the child's arm muscles.

Q5. At a swimming pool, a person of mass 55 kg jumps off a high diving board into the pool 4.5 m below the board.
(a) Calculate the loss of potential energy of the person due to this vertical descent of 4.5 m,
(b) Describe the energy changes of the person after leaving the diving board.

Efficiency and power

- **The power output** of a machine is the energy it supplies each second to the object it drives.
- **The power input** to a machine is the energy supplied each second to the machine.

Each second, a certain amount of energy is supplied to the machine and the machine delivers a certain amount of energy to whatever it is driving. The energy delivered per second by the machine cannot exceed the energy supplied to the machine each second because energy cannot be created in the machine. If all the energy delivered by the machine each second is equal to the energy supplied to it each second, the machine is perfectly efficient. In practice, friction is usually present between the moving parts of a machine so most machines are not perfectly efficient.

The **efficiency** of a machine is defined as

$$\frac{\textit{the energy per second delivered by the machine}}{\textit{the energy supplied to the machine per second}}$$

Because the energy delivered per second by the machine is its power output and the energy supplied to it per second is its power input, the *efficiency* = $\frac{\textit{power output}}{\textit{power input}}$

The maximum value of the efficiency of a machine is 1 as this is when the power output is equal to the power input so no power is wasted in the machine.

Note that the percentage efficiency of a machine is its efficiency × 100%.

Worked example

A 500 W electric winch raises a weight of 200 N at a speed of 0.4 m/s. Calculate
(a) the electrical energy supplied to the winch each second,
(b) the gain of potential energy of the weight in 1 second,
(c) the percentage efficiency of the winch.

$g = 9.8\ \text{m/s}^2$

Solution

(a) The power input to the winch = 500 watts. Therefore, in 1 second, the winch is supplied with 500 J of electrical energy.
(b) The weight gains a height of 0.4 m in 1 second. Hence the gain of potential energy in 1 second = weight × height gain in 1 second = 200 N × 0.4 m = 80 J.
(c) Efficiency = $\frac{\textit{energy delivered by machine each second}}{\textit{energy supplied to the machine each second}} = \frac{80\ \text{J}}{500\ \text{J}} = 0.16$

Percentage efficiency = 0.16 × 100% = 16%.

More about kinetic energy

Kinetic energy is the energy possessed by a moving object due to its motion. Consider an object of mass m which accelerates from rest at constant acceleration a for a certain time t.

- The speed of the object after time t, υ = acceleration × time = at
- Hence the distance moved, s = average speed × time = $\frac{1}{2}\upsilon t$ since the average speed = $\frac{1}{2}\upsilon$
- Therefore the work done on the object = force × distance moved

= (mass × acceleration) × distance moved

$= m\,a\,s = \mathrm{m}\,(\upsilon / t) \times \tfrac{1}{2}\,\upsilon\,t = \tfrac{1}{2}\,m\,\upsilon^2$

Because all the work done goes to kinetic energy, it follows that the kinetic energy of the object at speed υ is equal to $\tfrac{1}{2}\,m\,\upsilon^2$.

$$\textbf{Kinetic energy} = \tfrac{1}{2}\,m\,\upsilon^2$$

Worked example

Calculate the kinetic energy of a vehicle of mass 800 kg moving at a speed of 30 m/s.

Solution

Kinetic energy $= \tfrac{1}{2}\,m\,\upsilon^2 = \tfrac{1}{2} \times 800 \times 30^2 = 360\,000$ J

More about potential energy

The weight of an object of mass $m = mg$. To raise the object an equal and opposite force must be applied to it. Therefore if the object is raised through a height H, its potential energy increases by an amount equal to its weight × the gain of height which is equal to mgH. The same formula is used for loss of potential energy with H as the loss of height.

$$\textit{Change of potential energy} = mg\,H$$

Worked example $g = 9.8$ m/s^2

A ball of mass 0.5 kg is released from rest at a height of 3.0 m above a floor. Calculate

(a) the loss of potential energy of the ball just before it hits the floor

(b) its speed just before it hits the floor, assuming its kinetic energy is equal to the loss of potential energy.

Solution

(a) Loss of potential energy $= mgH = 0.5 \times 9.8 \times 3.0 = 14.7$ J

(b) Its kinetic energy just before hitting the floor = 14.7 J

$\therefore \tfrac{1}{2}\,m\,\upsilon^2 = 14.7$ where υ is its speed just before impact.

$$\text{Hence } \tfrac{1}{2} \times 0.5 \times \upsilon^2 = 14.7$$

$$\upsilon^2 = \frac{2 \times 14.7}{0.5} = 59$$

$$\upsilon = 7.7 \text{ m/s}$$

Summary

The efficiency of a machine

$$= \frac{\text{the energy per second delivered by the machine}}{\text{the energy supplied to the machine per second}} = \frac{\text{power output}}{\text{power input}}$$

Kinetic energy = $\frac{1}{2} m v^2$

Change of potential energy = $mg\,H$

Questions $g = 9.8$ m/s^2

Q6. A 500 watt electric winch raises a weight of 400 N through a height of 2.5 m in 10 seconds. Calculate (a) the gain of potential energy of the weight, (b) the electrical energy supplied to the winch, (c) the percentage efficiency of the winch.

Q7. A lift and its passengers have a total mass of 2800 kg. The lift takes 40 seconds to ascend a vertical distance of 20 m.
(a) Calculate the gain of potential energy per second of the lift.
(b) A 25 kW electric motor is used to raise the lift. Calculate the efficiency of the motor when it is used to make the lift and its passengers ascend.

Q8. An elevator in a factory raises boxes of weight 200 N through a height of 5.0 m. When the elevator is operating at normal speed, it transports three boxes every minute from the bottom to the top of the elevator. Calculate
(a) the gain of potential energy of a single box on the elevator,
(b) the power output of the elevator motor.

Q9. (a) Calculate the kinetic energy of
(i) a 60 kg athlete running at a speed of 10 m/s,
(ii) a 2000 kg truck moving at a speed of 30 m/s,
(iii) a ball of mass 0.2 kg moving at a speed of 20 m/s.

(b) Calculate the change of potential energy of
(i) a 50 kg person who jumps off a wall of height 0.80 m,
(ii) a ball of mass 0.2 kg thrown 12 m vertically into the air,
(iii) a 500 N weight raised through a height of 2.5 m.

Q10. At a fairground 'water chute', a small boat and its passengers descend through a height of 3.0 m on the chute. The boat reaches a speed of 5 m/s at the end of the chute. The total mass of the boat and its passengers is 300 kg. Calculate

(a) the loss of potential energy of the boat and its passengers,

(b) the kinetic energy of the boat at the end of the chute.

5 THERMAL PHYSICS

Machines at work and at home do many jobs that would otherwise be tiring or time-consuming. The production of iron and steel and the invention of the steam engine started the industrial revolution of the nineteenth century. In the twentieth century, the internal combustion engine and the jet engine took over from the steam engine, making travel to any place on Earth possible. In this chapter, we will look at the measurement of temperature, heat transfer and some of the thermal properties of materials in preparation for a deeper look in the next chapter at the principles behind energy transformations by heat engines.

Heat and temperature

In winter in Siberia, the temperature outdoors can fall below –40 °C. In summer in the Sahara desert, the temperature can rise above 40 °C. These temperatures are expressed on the **celsius scale**, denoted by the symbol °C (degree C). This scale is defined in terms of two 'fixed points' which are

- ice point, the melting point of pure ice at atmospheric pressure, 0 °C,
- steam point, the boiling point of pure water at atmospheric pressure, 100 °C.

A fixed point is a standard 'degree of hotness' which can be reproduced as required. A thermometer calibrated in °C would read 0 when in pure melting ice at atmospheric pressure and 100 when in steam at atmospheric pressure. Figure 5A shows a mercury-in-glass thermometer. The temperature of the mercury in the bulb is measured by reading where the end of the mercury column is against the scale. The thermometer is used by placing the bulb at the location where the temperature is to be measured then reading the scale after allowing sufficient time for the reading to stop changing.

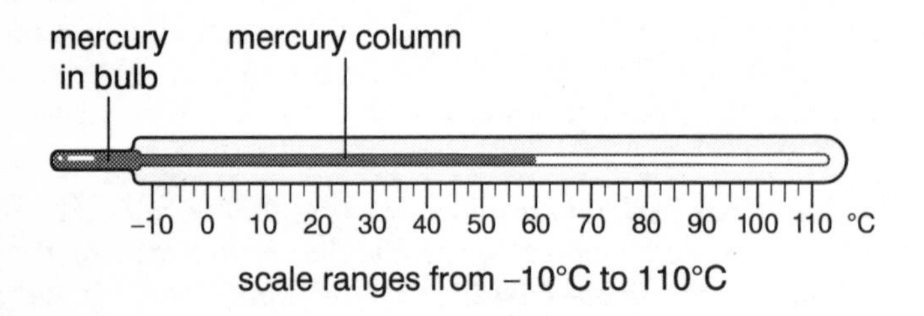

Figure 5A A mercury-in-glass thermometer

The mercury thermometer works because the mercury in the bulb expands when it becomes warmer and so it moves along the air column in the glass tube. Mercury is a very suitable liquid for use in a thermometer because it expands evenly with increasing temperature. Coloured alcohol is used as the liquid in thermometers for low temperature measurements because mercury freezes below –39 °C.

All thermometers give the same reading at ice point and at steam point because they are calibrated to read 0 °C and 100 °C respectively, at these two fixed points. Because the expansion of a liquid with increasing temperature varies from one liquid to another, the *gas thermometer* is chosen as the standard for readings between the fixed points. This type of thermometer consists of a gas-filled sealed bulb connected to a pressure gauge. The pressure of a gas in a sealed bulb increases if the gas temperature is raised and decreases if the temperature is lowered. The ratio of the pressure at steam point to the pressure at ice point is always the same, regardless of the amount or type of gas used. This is why the gas thermometer is used as the standard thermometer. All other thermometers are thus calibrated against a gas thermometer below ice point, between ice point and steam point and above steam point.

Absolute zero

An object placed in a refrigerator is made colder by the refrigerator because the refrigerator removes energy from the object. How cold can an object be? The lowest temperature possible is called **absolute zero**. No energy can be obtained from an object at this temperature. The pressure of a gas would be zero at absolute zero, provided the gas did not liquefy as it was cooled. By measuring the pressure of a fixed amount of gas at steam point then at ice point, the temperature for zero pressure (i.e. absolute zero) can be estimated. This temperature is –273 °C. No matter which gas or how much gas is used, the same value is always obtained for absolute zero. Here is an example of how this value is obtained from the pressure readings at ice point and at steam point. The readings are in kilopascals (kPa) which are explained on p. 154.

Steam point pressure reading = 250 kPa,

Ice point pressure reading = 183 kPa,

∴ a pressure change of 67 kPa (= 250 –183 kPa) is caused by a temperature change of 100 °C (from ice point to steam point), thus a pressure change of 1 kPa requires a temperature change of 1.49 °C (= 100 / 67).

To reduce the pressure from 183 kPa to zero would therefore require the temperature to be lowered by 273 °C (= 183 × 1.49) from ice point. Hence absolute zero = –273 °C. Precise measurements give –273.15 °C for absolute zero.

The ideas about absolute zero were developed by Lord Kelvin in the nineteenth century. **The absolute scale of temperature in kelvins** (K) is defined from two fixed points which are:

- absolute zero at zero kelvins (= –273 °C)
- the temperature at which ice, water and water vapour coexist (= 0 °C).

Hence absolute temperature in kelvins = temperature in °C + 273

The coldest place in the Universe

Low temperature research laboratories are the coldest places in the Universe. Scientists measure ultra-low temperatures in microkelvins (millionths of a kelvin). Methods to achieve such low temperatures involve making a suitably cold object work to reduce its energy. Low temperature research has produced some astonishing discoveries, including superconductors and superfluids.

Heat transfer

Heat is energy transferred due to a temperature difference. If two objects in contact with each other are at the same temperature, no heat transfer between the two objects takes place. The two objects are said to be in thermal equilibrium because there is no heat transfer between them. 'The baby in the bath' rule is based on the idea of thermal equilibrium; the parent holds the baby and dips his or her elbow (not the baby's elbow!) in the bathwater to test the water temperature. If the water temperature is the same as the elbow temperature, the baby can then be dipped in the water because he or she is in thermal equilibrium with the parent.

Heat transfer can occur in three different ways:

1 *Thermal conduction* takes place in solids, liquids and gases when there is a temperature difference between different parts of the substance. Metals are the best conductors of heat for the same reason that they are good conductors of electricity; they contain tiny particles called **electrons** which can move about carrying energy from one part of the metal to another part when there is a voltage or a temperature difference across the metal. We will meet the electron many times in this book.

  Insulating materials such as polystyrene sheets are made of substances in which all the electrons are trapped inside the atoms of the substance and cannot move about inside the substance. The presence of air pockets in an insulating material improves its insulation properties.

2 *Thermal convection* is the process of circulation that takes place in liquids and gases when there is a temperature difference between different parts of the liquid or gas. The density of a

liquid or gas varies with temperature so temperature differences in a liquid or gas causes density variations. Gravity makes more dense regions sink and less dense regions rise thus causing circulation currents. A hot air balloon rises because the air in the balloon is heated by a burner so becomes less dense and is therefore forced upwards. Vehicle cooling systems transfer heat from the engine to a radiator as a result of circulation of water heated by the engine and cooled in the radiator. The radiator is designed to lose heat to the surroundings by convection of air, assisted by a fan if necessary.

3 *Thermal radiation* is emitted from every surface. The hotter the surface is, the greater the thermal radiation from it. No substance is needed to carry the radiation as it can travel through a vacuum.

- A black surface is a much more effective absorber of thermal radiation than a shiny silvered surface is. You can notice this if you hold a piece of black paper and a sheet of tinfoil at the same distance from a hot object. The black paper warms up much faster than the tinfoil because it absorbs radiation from the hot object much more effectively.
- A black surface is a much more effective radiator of thermal radiation than a shiny silvered surface. If you want to keep an object warm, wrap it in tinfoil so it does not lose energy due to thermal radiation. If it is wrapped in black paper, it will cool down very quickly. A vehicle radiator is usually painted black to enable it to lose heat by radiation as well as by convection.

How to reduce your fuel bills

Heat flow is measured in watts where 1 watt is equal to a rate of energy transfer of 1 joule per second. A 1000 watt (= 1 kilowatt) heater supplies heat to its surroundings at a rate of 1000 joules per second. Fuel usage on fuel bills is usually priced in **kilowatt hours** (kW h) regardless of what the fuel is used for (e.g. heating, lighting, running a computer). One kilowatt hour is the energy used by a 1 kilowatt electric heater in 1 hour. For example, a gas bill that shows the fuel used as 11 500 kW h tells you that the amount of gas used gave the same energy as a 1 kilowatt heater for 11 500 hours. A fuel bill usually states the price per kilowatt hour of the fuel. For example, the cost of 11 500 kW h of fuel at a price of 1.3 p per kW h would be £149.50. Look at a recent fuel bill and work out how much fuel you use in your home each day and what the cost is.

Home heating bills can be reduced by fitting loft insulation and double glazing to reduce heat loss from indoors to outdoors. The picture of the house below shows these and other measures that can be taken to reduce fuel bills.

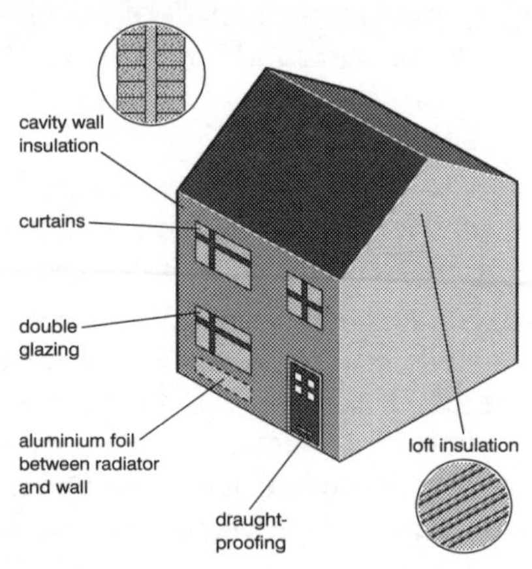

Figure 5B Reducing home heating bills

To work out if an installation such as double glazing is cost effective, the cost of the installation needs to be compared with an estimate of the amount that would be saved on fuel bills. Such an estimate can be made if the **U-value** of the installation is known. This is the heat flow (i.e. heat transfer per second) per square metre for a temperature difference of 1 °C.

- Loft insulation has a U-value of about 0.5 watts per square metre per °C. This means that the heat flow through 1 square metre of loft insulation is 0.5 watts for a temperature difference of 1 °C. The heat flow through 50 square metres of this loft insulation would be 250 watts for a temperature difference of 10 °C. The U-value of a 50 m^2 roof without loft insulation is about four times greater thus a temperature difference of 10 °C would cause 1000 watts of heat loss. Fitting loft insulation would reduce the heat loss by 750 watts or 0.75 kilowatts so in 90 days, the saving would amount to over 1600 kW h (= 0.75 × 24 hours / day × 90 days). At 1.3 p per kW h, the householder with loft insulation would therefore save about £21. Clearly, the calculation assumes an average temperature difference of 10 °C.

- Double glazing reduces heat loss due to conduction because a layer of air is trapped between two glass panes. Air is a much better insulator than glass and therefore a double glazed window cuts heat loss by a factor that depends on the thickness of the layer of trapped air. For a temperature difference of 10 °C, a typical single pane window loses heat at a rate of about 50 watts per square metre whereas a double glazed window loses heat at a much smaller rate of about 30 watts per square metre.
- Cavity wall insulation consists of plastic foam which is pumped into the cavity and which then sets inside the cavity. The foam provides a very effective insulating layer in the cavity between the two brick walls. A typical cavity wall without insulation in the cavity loses heat at a rate of about 15 watts per square metre for a 10 °C temperature difference; cavity wall insulation can reduce this by a factor of three to about 5 watts per square metre.
- Other less expensive measures to reduce heating bills include:
  1. draught-proofing of doors to cut down on heat loss due to convection,
  2. placing aluminium foil behind any radiator attached to an outside wall to cut down on thermal radiation from the radiator to the wall,
  3. closing the curtains (at night!) to cut down on heat loss due to radiation through the window.

Summary

Absolute temperature in kelvins = temperature in °C + 273

Heat is energy transferred due to a temperature difference.

Thermal conduction is due to energy transfer between particles in a substance. Thermal convection occurs in liquids and gases and is due to circulation. Thermal radiation is radiation emitted by a surface due to its temperature.

The U-value of an insulator is the heat flow (i.e. heat transfer per second) per square metre for a temperature difference of 1 °C.

Questions

Q1. State the temperature of (a) ice point, (b) steam point, (c) absolute zero on
(i) the celsius scale, (ii) the absolute scale.

Q2. Explain why (i) a saucepan has a plastic or wooden handle, (ii) the outlet pipe from a hot water tank is connected to the top of the tank, (iii) a hot potato in tinfoil stays hot much longer than an unwrapped hot potato.

Q3. (a) A semi-detached house (Number 1) has an uninsulated roof of area 60 m^2, single pane windows of total area 10 m^2 and external walls without cavity insulation of total area 120 m^2. Copy and fill in the table below to estimate the heat loss throught the roof, windows and walls for a 10 °C temperature difference.
(b) The adjoining semi-detached house (Number 3) has loft insulation, double glazing and cavity wall insulation. Complete the table below to estimate the heat loss from this house for a 10 °C temperature difference.
(c) Both houses are heated by gas-fired central heating at a cost of 1.3 p per kW h. Estimate the difference between the cost of gas in one day for these two houses.

Temperature difference = 10°C	U-value /W m^{-2} °C^{-1}	Area / m^2	Number 1's rate of heat loss / W	Number 3's rate of heat loss / W
Uninsulated roof	2.0	60		—
Insulated roof	0.5	60	—	
Single pane window	5.0	10		—
Double glazed window	3.0	10	—	
External wall with no cavity insulation	1.5	120		—
External wall with cavity insulation	0.5	120	—	
Total rate of heat loss	—	—		
Total cost per day at 1.3 p / kW h	—	—		

Thermal properties of materials

Metals heat up much more easily than most non-metals which is why the interior of a parked vehicle can become very hot in summer. The energy possessed by a substance due to its temperature is referred to as *thermal energy*. All substances are composed of molecules which are the smallest particles of a substance. In solids, the molecules are held together rigidly by strong bonds. In liquids, the bonds are not strong enough to keep the molecules in a rigid structure. In a gas or a vapour, the molecules move about freely at large distances from each other. See p. 150 for more information. When energy is supplied to a substance to increase its thermal energy, the particles of the substance

- gain kinetic energy if the temperature of the substance increases,
- use the energy supplied to break the bonds between the molecules, if the substance changes state from a solid to a liquid or to a gas or from a liquid to a gas.

Melting points and boiling points

When a pure solid is heated and heated, its temperature increases until it reaches the *melting point* of the solid. The solid changes to a liquid at this temperature, provided it continues to be heated. Continued heating raises the temperature of the liquid to *boiling point*, the temperature at which the liquid boils. The melting point of a solid is characteristic of the solid and the boiling point of a liquid is characteristic of the liquid and therefore can be used to identify the substance. Freezing is the reverse process to melting and it occurs when the liquid has been cooled to its freezing point which is the same temperature as the melting point. For example, the freezing point of water and the melting point of pure ice is 0 °C. The reverse process to boiling is condensation. If a gas in a container is cooled sufficiently, liquid forms on the surface of the container due to condensation.

Winter journeys

Roads in winter are usually covered in grit and salt if freezing weather is due. Icy roads are dangerous because vehicles are difficult to control on ice due to lack of grip on the road surface. Salt dissolves in water and salty water has a lower freezing point than pure water. If the freeze is severe, grit particles in the ice help to provide some grip. Nevertheless, vehicle speeds must be greatly reduced in freezing weather conditions to reduce the risk of traffic accidents.

Specific heat capacity

The specific heat capacity of a material is the energy needed to raise the temperature of unit mass of material by one degree. For example, the specific heat capacity of water is 4200 joules per kilogram per 0 °C. This means that

- 4200 J of energy must be supplied to raise the temperature of 1 kilogram of water by 1 °C,
- 42 000 J of energy must be supplied to raise the temperature of 1 kilogram of water by 10 °C,
- 84000 J of energy must be supplied to raise the temperature of 2 kilograms of water by 10 °C.

More generally, to raise the temperature of mass m of a substance from T_1 to T_2,

$$\textit{the energy needed } \Delta E = mc\,(T_2 - T_1),$$

where c is the specific heat capacity of the material. The unit of c is $J\,kg^{-1}\,{}^0C^{-1}$.

Notes

1 *To calculate c using the formula $\Delta E = mc(T_2 - T_1)$, rearrange the formula to make c the subject as follows:*

$$mc\,(T_2 - T_1) = \Delta E$$

$$\therefore c = \frac{\Delta E}{m\,(T_2 - T_1)}$$

2 *Specific heat capacities of some other substances are listed in the table opposite.*

Substance	Specific heat capacity in J kg^{-1} $^{\circ}C^{-1}$
Water	4,200
Copper	380
Aluminium	900
Ice	2,100

How to measure the specific heat capacity of a metal

You will need a thermometer, a 12 V power supply, a joulemeter, a 12 V electric heater, an insulated metal block with slots for the heater and the thermometer and a balance.

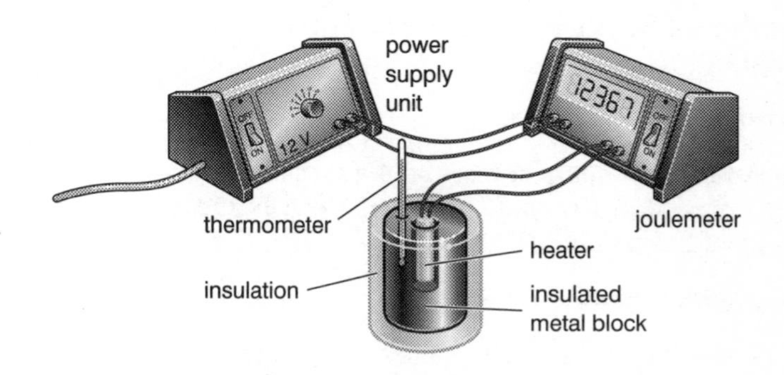

Figure 5C Measuring the specific heat capacity of a metal

1. Use the balance to measure the mass m of the metal object to be tested and write your measurement down.
2. Connect the heater to the joulemeter and connect the joulemeter to the 12 V power supply but do not switch the power supply on yet. Place the heater and the thermometer in the slots in the block. Measure and note the initial temperature T_1 of the block and the initial reading of the joulemeter.

3 Switch the power supply on for five minutes. Note the joulemeter reading at the end of this time and the temperature T_2 of the block.

4 Calculate the energy supplied, ΔE, in joules from the difference of the joulemeter readings and calculate the temperature difference $(T_2 – T_1)$.

5 Use the formula $\Delta E = mc(T_2 – T_1)$ to calculate the specific heat capacity, c, of the block.

Example of measurements for an aluminium block:

Mass of block, m = 1.0 kg

Initial temperature, T_1 *= 15 °C*

Highest temperature, T_2 *= 42 °C*

Initial reading of the joulemeter = 22 500 J

Final reading of the joulemeter = 46 800 J

Show that the results give the specific heat capacity of aluminium = 900 J kg^{-1} *°C*

Summary

To raise the temperature of mass m of a substance from T_1 to T_2, the energy needed $\Delta E = mc(T_2 – T_1)$, where c is the specific heat capacity of the material. The unit of c is J kg^{-1} °C^{-1}.

Questions

Use the specific heat capacity values given in the table on p. 71 in the questions below.

Q4. (a) Calculate the energy needed to heat 1.5 kg of water from 10 °C to 100 °C.

(b) Calculate the energy needed to heat an aluminium kettle of mass 0.4 kg from 10 °C to 100 °C.

(c) Calculate the energy needed to heat 1.5 kg of water in an aluminium kettle of mass 0.4 kg from 10 °C to 100 °C.

Q5. (a) Calculate the final temperature when 2500 J of thermal energy is supplied to a copper block of mass 0.51 kg at an initial temperature of 12 °C.

(b) An insulated aluminium block of mass 2.0 kg is heated by a low voltage 36 W electric heater for 10 minutes. Calculate

(i) the energy supplied by the heater,
(ii) the temperature rise of the aluminum block.

Q6. An insulated copper hot water tank has a mass of 20 kg and contains 25 kg of water at 15 °C.
(a) Calculate the energy needed to heat the tank and the water to 45 °C.
(b) Calculate how long a 3.0 kW electric immersion heater in the tank would take to supply the energy calculated in (a).

Specific latent heat

Energy must be supplied to a solid to melt it. There is no temperature rise at the melting point because the energy supplied is used by the molecules in the solid state to break the bonds between them. Energy must be supplied to a liquid at its boiling point to boil the liquid away. The energy supplied is used by the molecules in the liquid state to break the bonds between them. Because there is no temperature rise at the melting point or the boiling point, the energy supplied is called **latent heat**. The word 'latent' means 'hidden'.

The specific latent heat of a solid or a liquid substance for a given change of state is the energy needed to change the state of unit mass of material, without change of temperature. For example, the specific latent heat of fusion of ice is 336,000 joules per kilogram. This means that

- 336 000 joules of energy is needed to melt 1 kilogram of ice,
- 672 000 joules of energy is needed to melt 2 kilograms of ice,
- 1.68 million joules of energy is needed to melt 5 kilograms of ice.

More generally, to change the state of mass m of a substance at constant temperature,

$$\text{the energy needed } \Delta E = m\,l,$$

where l is the specific latent heat of fusion (for melting or solidifying) or vaporization (for vaporizing, boiling or condensing) or sublimation (for a solid which vaporizes directly or a vapour which forms a solid without liquefying first) for that substance. The unit of l is $\mathrm{J\,kg^{-1}}$.

Note: to calculate l using the formula $\Delta E = m\,l$, rearrange the formula to make l the subject as follows:

$$m\,l = \Delta E$$

$$\therefore l = \frac{\Delta E}{m}$$

How to measure the specific latent heat of steam

Use an electric kettle with a known power rating in watts. Fill an electric kettle with a measured volume of water. Switch the kettle on and boil the water in it for 2 minutes. You will need to open some windows when you do this to prevent condensation in the room. After 2 minutes, switch the kettle off and let the water cool. When it is sufficiently cool, measure the volume of the water still in the kettle. The difference between the volume of water at the start and the volume at the end is the volume of water boiled away. Hence calculate the mass of water boiled away, given the density of water is 1.0 kilogram per litre.

Calculate the energy supplied to the kettle from the power rating in watts × 300 seconds.

Hence determine the latent heat of steam from the energy supplied and the mass of water boiled away.

Sample measurements

Initial volume of water = 1.50 litres

Final volume of water = 1.35 litres

Power of kettle = 3000 watts

Energy supplied in 2 minutes = 3000 watts × 120 seconds = 360,000 joules

Volume of water boiled away = 0.15 litres

Mass of water boiled away = density of water × volume boiled away = 1.0 kg / litre × 0.15 litres = 0.15 kilograms,

∴ specific latent heat of steam = 360 000 joules / 0.15 kilograms = 240 000 joules per kilogram.

Summary

To change the state of mass m of a substance at constant temperature, the energy needed $\Delta E = m\,l$, where l is the specific latent heat of fusion (for melting or solidifying) or vaporization (for vaporizing, boiling or condensing) or sublimation (for a solid which vaporizes directly or a vapour which forms a solid without liquefying first) for that substance. The unit of l is $J\,kg^{-1}$.

Questions

The specific latent heat of ice is 336 000 joules per kilogram. The specific latent heat of steam is 2.3 million joules per kilogram. The specific heat capacity of water is 4200 joules per kilogram per °C.

Q7. (a) Calculate the energy needed to boil away 0.5 kg of water.
(b) Calculate the mass of ice that could be melted, given the amount of energy calculated in (a).

Q8. On a cold sunny day, a bucket containing 2.5 kg of ice at 0 °C melts and warms to a temperature of 5 °C.
(a) Calculate the energy needed to melt (i) 2.5 kg of ice at 0 °C, (ii) heat the melted ice to 5 °C.
(b) The bucket of ice takes 3.5 hours to melt and warm to 5 °C. Calculate the energy per second gained by the bucket.

Q9. A plastic beaker containing 0.20 kg of water at 15 °C placed in a freezer cools down and freezes within 20 minutes. Calculate
(a) the energy removed from the water to (i) cool it down to 0 °C, (ii) freeze it.
(b) the rate at which energy is removed from the water.

Q10. A 3000 watt electric kettle is used to boil water. Calculate
(a) the electrical energy supplied to the kettle in 1 minute,
(b) the mass of water boiled away in 1 minute. Assume all the energy supplied is used to boil the water.

6 ENGINES AND THERMODYNAMICS

The science of thermodynamics is about temperature, heat and work and it provides rules and limits on the use of energy. At the start of the nineteenth century, many scientists thought of heat as an invisible fluid, referred to as 'caloric'. Two decades later, the caloric theory had been consigned to oblivion, replaced by the new theory that heat is a form of energy. By the middle of the nineteenth century, scientists had worked out the laws of thermodynamics including the all-important second law which tells us that energy tends to spread out and become less useful as a result. In this chapter, we will look at these ideas and how they govern the efficiency limits of engines. In addition, we will make a survey of present and future energy supply and demand, including fuel supplies and renewable energy resources.

Heat engines

Before the invention of the steam engine, power to drive machinery was provided by windmills or watermills and power for transport was provided by horses on land and wind and ocean currents at sea. James Watt recognized the potential of steam power, as a result of observing jets of steam from a boiling kettle. The invention and development of the steam engine created industries that were able to produce and distribute goods and food on a much larger scale than was ever possible previously. Nowadays, the steam engine has faded into obscurity, replaced by engines such as the internal combustion engine, the diesel engine, the electric motor and the jet engine. Steam or gas turbines drive electricity generators in power stations to supply electricity to homes, industry and rail transport. Although an electric motor is not an engine as it does not have its own source of energy, the electricity for it is supplied by an electricity generator driven by a turbine in a power station. Thus we rely on engines of one form or another for transport and for our energy needs at home and at work. Most types of engines are heat engines as they obtain energy by burning fuel. An exception is a turbine in a hydroelectric power station where water flow from an upland reservoir drives the turbine as it flows downhill. The aerogenerator, a wind-driven electricity generator, is another exception.

Steam engines and turbines, internal combustion engines, jet engines and rocket engines are described as **heat engines** because each type of engine uses energy from a high temperature source to do work. In other words, a heat engine uses heat to do work.

To drive a **steam engine**, burning fuel is used to boil water in a boiler so producing steam at high pressure. The steam is fed via pipes and valves to cylinders, each containing a piston coupled to a drive shaft. The valves are opened and closed automatically by rods connected to the drive shaft. Thus the pressure of the steam drives the piston back and forth along the cylinder, rotating the drive shaft in the process.

In a four-stroke internal combustion engine, an electric spark is used to explode a mixture of gasoline and air, forcing the piston to move along the cylinder and turn the drive shaft. An inlet and an outlet valve are opened and closed automatically in a sequence of four piston strokes:

1. Squeeze stroke: inlet valve closed; outlet valve closed; the piston moves into the cylinder and compresses the fuel mixture in the cylinder. A spark explodes the mixture at maximum compression, causing the temperature and pressure of the air in the cylinder to shoot up.
2. Power stroke: inlet valve closed; outlet valve closed; the piston is forced out by the high pressure in the cylinder. The pressure in the cylinder drops as the piston moves out.
3. Sweep stroke: inlet valve closed; outlet valve opens; the piston returns and forces the combustion products out of the cylinder via the outlet valve.

4 Draw stroke: inlet valve open; outlet valve closed; the piston moves out of the cylinder (driven by the power stroke of one of the other cylinders) and draws more fuel and air into the cylinder for the next 'squeeze' stroke.

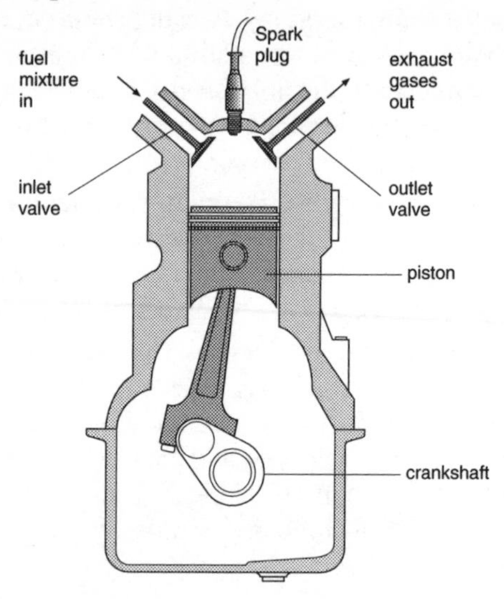

Figure 6A The internal combustion engine

In a **steam turbine**, steam jets are directed at the blades of a turbine wheel, causing the wheel to rotate. The wheel is usually one of several wheels on the same shaft inside an enclosed turbine block. The base of the block is kept cool by means of water pumped through pipes in the block.

In a **jet engine**, air is drawn into the engine and fuel from the feed pipes enters the air stream. The fuel mixture is burned downstream continuously in the burner chamber, thus raising the air pressure and temperature enormously. The hot air emerges at high speed through the outlet duct into the atmosphere, thus forcing the engine forward.

Heat transfer requires a low temperature 'sink' as well as a high temperature source. In the examples above, the atmosphere provides the low temperature sink, either directly or indirectly via a cooling system. Not all the heat from the high temperature source can be used by the engine to do work. This restriction is because the low temperature sink must accept some of the energy transferred from the high temperature source, otherwise the engine would overheat and stop working. For example, in a steam turbine, steam jets at high pressure are used to drive the turbine wheel; the turbine is kept cool by means of water pumped through pipes that pass through the engine block. If this cooling system was switched off, the engine temperature would rise. The pressure inside the engine would therefore rise and this would lessen and eventually cancel the force of the steam jets that drive the turbine wheels.

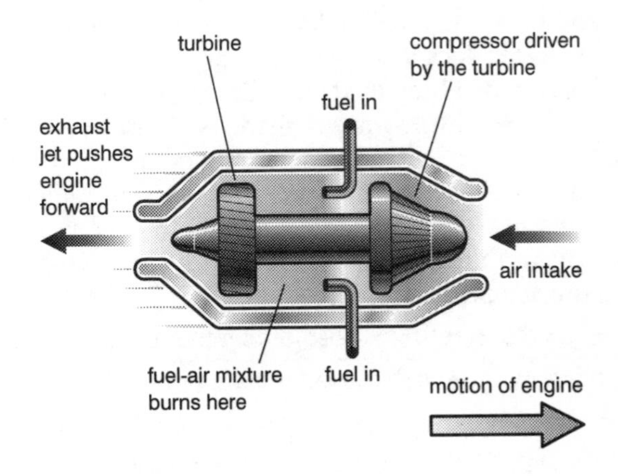

Figure 6B The jet engine

The efficiency of a heat engine

A heat engine does work as a result of using heat from the combustion of fuel. Some of the heat from the fuel is not used to do work and is supplied to the low temperature 'sink'. The energy used to do work is referred to as useful energy. The energy supplied to the low temperature 'sink' is wasted by the engine as it is not used to do work. The efficiency of a heat engine is the fraction of the energy supplied to it from the fuel which is used to do work. For example, an engine with an efficiency of

- 0.2 converts 20% of the energy from its fuel into useful energy and wastes 80% of the energy from the fuel.
- 0.5 converts half the energy from its fuel into useful energy and the other half is wasted,
- 0.8 converts 80% of the energy from its fuel into useful energy and wastes 20% of the energy from the fuel.

The principles of heat engines were established in the early nineteenth century by Sadi Carnot, a French scientist. Carnot realized that the lower the temperature of the 'sink', the more efficient the engine would be. This idea is like a hydroelectric scheme where water from an upland lake flows through pipes into a generator and then into a reservoir. The lower the reservoir, the more work the water does as it passes through the generator. The overall efficiency depends on the height drop from the upland lake to the generator. However, just as there is a limit to the drop between the upland lake and the generator, there is a limit to how cold the heat sink could be, namely the lowest possible temperature which is absolute zero. Carnot worked out that for any heat engine operating at maximum efficiency between a high temperature source and a low temperature sink,

$$\frac{\text{the heat supplied to the sink}}{\text{the heat supplied by the source}} = \frac{\text{absolute temperature of the sink}}{\text{absolute temperature of the source}}$$

For example, if the engine was operating between 1000 K and 300 K,

$$\frac{\text{the heat supplied to the sink}}{\text{the heat supplied by the source}} = \frac{300}{1000} = 0.30$$

In this example, for every 100 joules supplied from the source, 30 joules would be taken by the sink. The work done by the engine in this situation would therefore be 70 joules from every 100 joules supplied by the source. In other words, the efficiency of this heat engine is 0.7.

Carnot proved that for any heat engine operating between a source at temperature T_1, and a sink at lower temperature T_2,

$$\textit{the maximum efficiency of such an engine} = \frac{(T_1 - T_2)}{T_1}$$

Thus an internal combustion engine in which the air is heated to 700 K (when the fuel mixture is ignited) and cooled to 400 K has a maximum efficiency of 0.57 (= (700 – 400) / 700).

If combustion of fuel in this engine releases energy at a rate of 50 kilowatts (= 50 000 joules per second), the maximum power output of the engine would be 28.5 kW (= 50 kW × 0.57) and the energy per second wasted would be 21.5 kW (= 50 kW – 28.5 kW).

The laws of thermodynamics

Carnot's ideas about heat engines were developed into the laws of thermodynamics which are the principles behind energy transformations involving heat.

1 *The First Law of Thermodynamics* states that when an object or a system of objects does work or gains heat, the change of internal energy of the object(s) is equal to the difference between the heat gained and the work done by the object(s).

- *Work* is energy transferred by means of a force when the force moves its point of application in the direction of the force. See p. 51.
- *Heat* is energy transferred due to a difference of temperature. The relationship between heat and temperature difference is a bit like the 'chicken and egg question'; which one comes first? A difference of temperature is said to exist whenever energy is transferred by means other than work. Heat is energy transferred by any means other than work. See p. 64.
- *The internal energy of an object* is the energy it possesses, regardless of its speed or position. For example, if the temperature of a solid object is increased by heating it, the internal energy of the solid increases as its atoms vibrate more.

The First Law of Thermodynamics follows from the Principle of Conservation of Energy as heat and work are the means by which an object gains or loses energy. The work done per second by a heat engine is equal to the difference between the heat transfer per second *from* the high temperature source and the heat transfer per second *to* the low temperature sink. An engine doing work W does so as a result of accepting heat Q_1 from a high temperature source and rejecting heat Q_2 to a low temperature sink. Thus $Q_2 = Q_1 - W$.

2 *The Second Law of Thermodynamics* states that energy tends to spread out and become less useful whenever it is transferred between objects in an isolated system.

For example, when an electric winch is used to raise a weight, energy is transferred from the electrical supply to the weight. However, some of the energy supplied is wasted due to friction between the moving parts and due to the heating effect of the electric current in the circuit. This wasted energy cannot be recovered and used to do work as it is dissipated by means of heat transfer to the surroundings. The energy from the electrical supply gained by the weight is an example of useful energy. The energy from the electrical supply not gained by the weight is wasted as heat transfer to the surroundings.

For a heat engine operating at its maximum efficiency, the work done by the engine must be less than the heat supplied from the high temperature source. As we saw on p. 78, this is because heat transfer from the high temperature source would stop if heat transfer to the low temperature sink was prevented. Some heat transfer to the low temperature sink must therefore take place and therefore some energy from the source is wasted and not used to do work.

Energy for the future

The problem with energy is that it tends to spread out. Even when energy is stored, as happens when a car battery is charged or when a weight is raised, some of the energy used in the storage process is wasted due to friction or resistance and cannot be recovered. Although energy cannot be created or destroyed, it tends to spread out when it is used. Because our energy resources are finite, such resources will eventually run out. Each person in Europe and North America uses energy at an average rate of about 8000 joules per second which amounts to 250 000 million joules per person each year. For a world population of about 6000 million people (likely to exceed 8000 million by 2050), each using energy at a rate of 8000 joules per second, the rate of energy usage would therefore be about 1500 million million million joules each year. At present, world energy use is about 300 million million million joules per year or about 1500 joules per second per person. Global energy use will probably double by 2050 as there will be more people on the planet and each person will use more energy as living standards rise.

Fossil fuels such as coal, wood, oil and gas provide most of the energy we use at present. These fossil fuels are substances which release energy when burned in air. Fossil fuels are extracted from the Earth, having taken millions of years to form as a result of geological changes which have buried and compressed dead vegetation to form coal and dead marine life to form oil and gas. The energy released when fossil fuel is burned was stored millions of years ago by a living object from sunlight. Fuel is not renewable; once burned, the energy is released and the substance is changed into other compounds such as carbon dioxide and carbon monoxide. Each kilogram of fossil fuel releases about 30 million joules when burned. Approximately 10 million million kilograms of fossil fuel is burned each year to meet worldwide energy demands. The world reserves of fossil fuels are reckoned to be about 15 00 million million kilograms, sufficient to last for about 150 years. However, oil and gas which constitute about 30% of present fossil fuel reserves are being used at a faster rate than coal and will probably be used up within the next 50 years. Industrial countries will need to develop other energy resources soon or cut back to a 'low energy' future.

Nuclear reactors release energy from uranium contained in fuel rods in the reactor. The energy released is removed by a coolant which is pumped through sealed pipes passing through the reactor core. The hot coolant flows through a heat exchanger where it is used to raise steam before being pumped back cooler into the reactor core. The steam from the heat exchanger is used to drive turbines to produce electricity. A large nuclear reactor produces about 1500 megawatts of electrical power. See p. 193 for more details. Note that 1 megawatt (MW) = 1 million watts.

At present, about 8% of worldwide energy demand is met by nuclear power stations as they provide between a quarter and a third of the electricity supplies in industrial countries. World reserves of uranium will probably last about 50 years at the present rate of use. However, nuclear reactors at present use a form of uranium known as uranium 235 which is no more than about 2% of natural uranium. The other 98% of natural uranium is uranium 238 which is not used in the present type of nuclear reactors. The lifetime of the world's reserves of uranium would be extended many times if reactors called fast breeder reactors were used as the fuel for these reactors is plutonium which is made in the present generation of nuclear reactors from uranium 238. The used fuel rods from nuclear reactors are highly radioactive and remain so for thousands of years. Radioactivity is harmful to human health as it is known to cause cancer. For this reason, nuclear reactors currently in use may not be replaced when they reach the end of their designated working life. Electricity cut backs will probably happen unless alternative means of generating electricity on a large scale are developed. In Britain, electrical power demand varies from about 20 000 megawatts in summer to about 50 000 megawatts in winter.

Renewable energy resources such as hydroelectric power stations and windfarms of aerogenerators will need to be developed on a massive scale to meet our electricity demands when oil and gas reserves are used up if more nuclear reactors are not constructed. A renewable energy resource is a source of useful energy that does not require fuel. Most renewable energy resources use solar energy, either directly as in solar panels or indirectly via the Earth's atmosphere which is heated by the Sun. Wind turbines are driven by winds created in the atmosphere, wave generators are driven by waves created by winds, hydroelectric generators are driven by rainwater running downhill. Other renewable resources include geothermal energy which is obtained by pumping water into hot underground rock basins and then using the steam generated to drive turbines. Tidal power is another non-solar energy resource. The tides rise and fall twice each day because the Moon pulls on the Earth's oceans as the Earth spins. A tidal power station is a barrier which traps seawater at high tide and releases it through channels in the barrier to drive turbines and produce electricity.

■ Could solar energy be used, directly or indirectly, to meet our energy needs? On a clear day, each square metre of the Earth's surface receives 1400 joules every second from the Sun when the Sun is directly overhead. Taking account of cloud conditions and

Oil	Coal	Natural gas	Nuclear power	Hydroelectricity
40%	27%	22%	8%	3%
50 years	300 years	70 years	60 years*	indefinitely

[*for thermal reactors (3000 years for fast breeder reactors)]

(a) World Fuel Use

■ World fuel use is about 400×10^{18} J per year. This is presently met from energy sources as shown. The list also shows how many years present fuel reserves will last at the 1995 rate of use.

Oil	Coal	Natural gas	Nuclear power	Hydroelectricity
59%	13%	21%	7%	<0.5%
<50 years	1300 years*	40 years	imported	indefinitely

[*based on 1983 estimates if coalfields closed since 1983 are re-opened]

(b) UK Fuel Use and reserves

■ UK fuel reserves are shown. The total energy used by all sources in the United Kingdom is about 3.5% of the world fuel use. The list also shows the lifetime of present UK fuel reserves at the 1995 rate of use.

Figure 6C Energy resources

latitude, a **solar panel** fitted to a house roof in the northern hemisphere receives on average about 100 joules per second per square metre in the daytime. Each person would need about 80 square metres of solar panels to meet his or her energy needs. Solar panels could make a contribution to our energy needs but would need to cover an enormous area to have a significant effect. A solar heating panel fitted to a house roof uses sunlight to heat water that flows at a steady rate through the panel. The hot water collects in an insulated tank connected to the hot water taps in the house. Solar cell panels consist of **solar cells** that generate electricity directly from sunlight. Fuel for vehicles might in future be produced on a large scale by using solar cells to produce hydrogen gas from water, a process than involves no more than passing electricity through water. Vast arrays of solar cells covering large areas would still be necessary though.

Wind turbines are electricity generators on top of tall towers, each generator turned by propellors driven by the wind. A large wind turbine is capable of producing about a megawatt (= 1 million watts) of electrical power. A windfarm would need to have thousands of wind turbines to produce the same amount of power as a 5000 megawatt nuclear power station.

Hydroelectric power stations in hilly areas where there is plenty of rainfall contribute to the electricity supplies in some countries, mostly on a small scale at present. A single hydroelectric power station is not likely to produce much more than about 100 megawatts but a large number of such stations could make a significant contribution to our energy needs.

A **tidal power station** in a suitable coastal location could provide as much electricity as a large power station but it would need to cover a large area. A tidal area of 400 square kilometres could trap 4000 million cubic metres of water each high tide for a height difference of 10 metres between high tide and low tide. The density of water is 1000 kilograms per cubic metre so the mass of such a large volume of water is 4 million million kilograms. As explained on p. 58, this amount of water would release 80 million million joules of potential energy for a 2 metre drop of height. Over the 12 hours between successive high tides, the average rate of release of energy would therefore be about 2000 million joules per second, sufficient to generate 1000 megawatts of electrical power from electricity generators operating at 50% efficiency.

Wave generators floating offshore could also make a significant contribution to our energy needs. A wave generator is in two sections hinged together so they can move relative to each other, thus turning a generator in one of the sections. The incoming waves make the two sections rock relative to each other. Scientists reckon that about 1 kilometre of coastline could produce about 50 megawatts of electrical power.

Summary

For **a heat engine** operating between a source at temperature T_1, and a sink at lower temperature T_2, the maximum efficiency of such an engine = $\frac{(T_1 - T_2)}{T_1}$.

The First Law of Thermodynamics states that when an object or a system of objects does work or gains heat, the change of internal energy of the object(s) is equal to the difference between the heat gained and the work done by the object(s).

The Second Law of Thermodynamics states that energy tends to spread out and become less useful whenever it is transferred between objects in an isolated system.

Questions

Q1. For each type of heat engine listed below, what is (i) the source of the high temperature, (ii) the low temperature sink?
(a) An internal combustion engine,
(b) A jet engine,
(c) A steam engine.

Q2. Calculate the efficiency of an internal combustion engine in which air is heated to 650 K when the surrounding temperature is (a) 300 K, (b) 320 K.

Q3. Britain's electricity demand in winter is about 50 000 megawatts. About one-third of Britain's electricity is provided by oil and gas-fired power stations, about a third is provided by coal-fired power stations and about a third is produced by nuclear power stations.
(a) How many megawatts of electricity is provided by these nuclear power stations?
(b) How many megawatts of electricity is provided by fossil fuel power stations?
(c) State one advantage and one disadvantage of (i) a nuclear power station, (ii) a coal-fired power station, (iii) a gas-fired power station.

Q4. Britain's oil and gas reserves will probably run out by 2050. Suggest possible strategies that could be developed to make up the 'energy gap' after these reserves have been used up.

Q5. (a) A supplier of solar heating panels estimates that a panel of area 1 square metre absorbs 500 W of solar energy on a sunny day. Calculate how many of these panels would be needed to supply 3000 W of solar energy on such a day.

(b) A wind turbine is capable of supplying 2 MW of electrical power. How many such wind turbines would be needed to supply the same amount of electrical power as a 5000 MW nuclear power station?

(c) Wave generators could provide 50 MW of electrical power for every kilometre length of coastline. What length of coastline would need to be used to generate the same amount of electrical power as a 5000 MW nuclear power station?

7 ELECTRICITY

Electricity provides most of the power we use at home and in the workplace. Before the electricity distribution system was established, people used gas or solid fuels for heating and cooking. Lighting was provided by gas or oil lamps or candles. Electrical appliances are indispensable as anyone who has suffered an electrical power failure knows. In many underdeveloped countries, electricity does not reach remote villages and consequently most people live in poverty. The principles that underpin electricity generation and distribution were discovered in the nineteenth century and electricity distribution systems were set up in the late nineteenth century and the early decades of the twentieth century. In the future, renewable energy sources such as solar panels and wind turbines will probably provide electricity without causing pollution problems in urban regions as well as providing electricity in remote regions. In this chapter, we will look in depth at the principles of electricity and how electricity is generated and distributed.

The nature of electricity

Any material that allows electricity to pass through it is known as an *electrical conductor*. All metals and certain non-metals such as graphite conduct electricity. Materials that do not allow electricity to pass through are known as *electrical insulators*. Examples of electrical insulators include polythene, nylon, bakelite, air and oil. An electric torch consists of a battery connected to a torch bulb, usually with one terminal of the battery connected directly to the torch bulb and the other terminal connected to the torch bulb via a switch. When the switch is closed, the torch bulb lights because the two metal parts of the switch make contact. When the switch is opened, the two metal parts move apart and the bulb goes off. A complete circuit of electrical conductors is needed for electricity to pass through the torch bulb. If there is a gap in the circuit, no electricity can pass round the circuit.

Static electricity

Static electricity is produced in a thunderstorm when clouds become charged with electricity. When charged clouds can hold no more electricity, they discharge to Earth in massive lightning strokes. A less dramatic demonstration of static electricity can be achieved by rubbing an inflated balloon with a dry cloth. You can feel the static electricity discharging to your hand if you then touch the charged balloon. The charged balloon will attract bits of paper and will stick to a ceiling. Certain insulating materials such as glass, polythene, perspex, nylon and rubber also become charged when rubbed with a dry cloth.

A charged object exerts a force on any other charged object. For example, two charged polythene rods repel each other when they are held close together. This may be demonstrated by charging the end of one of the rods and then suspending it horizontally on a thread. The other rod is then charged at one end and repels the end of the suspended rod when placed close to it. The same effect is observed if two charged perspex rulers are held close together. However, if a charged perspex ruler is held near a charged polythene rod, the two objects attract each other. These tests show that

1 there are two types of electric charge,
2 two objects that carry like charge (i.e. the same type of charge) repel each other,
3 two objects that carry unlike charge (i.e. different types of charge) attract each other.

The two types of charge are referred to as *positive* and *negative* charge because they cancel each other out if one type is brought into contact with the other type. A charged polythene rod carries a negative charge. Therefore, any charged object that is repelled by a charged polythene rod carries a negative charge. A charged perspex ruler carries a positive charge and therefore it attracts a charged polythene rod.

The action of rubbing such a material charges it with electricity because tiny negatively charged particles transfer between the cloth and the material when they are rubbed. These charged particles are called **electrons** and they are in every atom of every substance. Every atom contains a positively charged nucleus which is surrounded by electrons. An uncharged atom has the same amount of negative charge as positive charge.

- If electrons are added to an uncharged atom, the atom becomes negatively charged because it now contains more negative charge than positive charge.
- If electrons are removed from an uncharged atom, the atom becomes positively charged because it now contains more positive charge than negative charge.

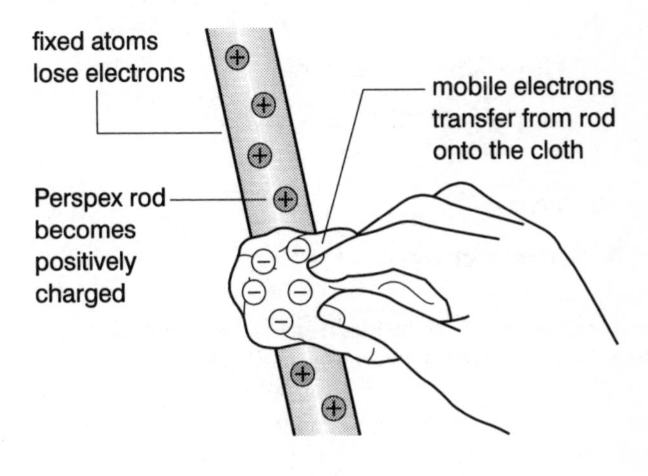

Figure 7A Charging by friction

Current and charge

An electric current is a flow of charge, usually carried by electrons. The electrons in an insulator are trapped inside the atoms but in a conductor some of the electrons are not trapped and can move about freely inside the conductor. A torch lamp circuit consists of a battery connected to the torch bulb via two wires and a switch. When the switch is closed, electrons pass round the circuit because all the parts of the circuit can conduct electricity. The battery forces the free electrons in the conductors one way round the circuit through the wires, the torch bulb, the switch and the battery.

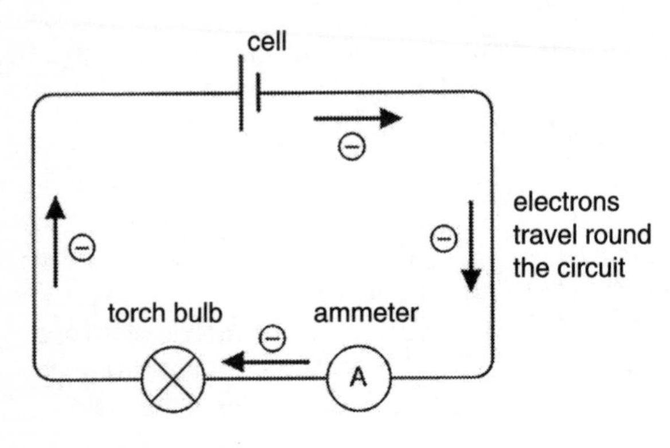

Figure 7A Electrons at work

When a circuit is switched on, electrons leave the negative terminal of the battery, move round the circuit and re-enter the battery at its positive terminal. The one-way flow of charge round an electric circuit was first deduced by the French scientist, Andre Ampere. Before Ampere's discovery, it was thought that positive charge flowed from the positive terminal of the battery and was cancelled out by negative charge flowing from the negative terminal. Ampere knew that an electric current along a wire caused a magnetic compass near the wire to deflect and he knew that this was because a magnetic field is created around the wire when an electric current was passed along the wire. He also noticed that the compass deflected in the opposite direction if the battery was reversed in the circuit. He realized that this observation could only be explained if only one type of charge flows round a circuit. Reversal of the battery caused the flow of charge in the circuit to reverse which reversed the direction of the magnetic field round the wire. However, he was not able to say if the charge is positive charge from the positive terminal or negative charge from the negative terminal of the battery. To prevent confusion, scientists agreed that the direction of current in a circuit should be the direction of flow of positive charge. We know now that an electric current in a circuit is a flow of electrons round the circuit from the negative terminal to the positive terminal of the battery. Nevertheless, the rule that the direction of current is the direction of flow of positive charge is still used today.

The unit of electric current is the *ampere* (A), defined in terms of its magnetic effect. A current of 1 A passing along a wire is due to 6.25 million million million electrons passing along the wire every second. Each electron carries the same amount of negative charge. The bigger the current in a wire, the greater the number of electrons passing along the wire each second. Therefore, the charge passing along a wire in a certain length of time is proportional to the current and the duration of the time interval.

Note: Small currents are measured in milliamperes (mA) or microamperes (μA), where 1 mA = 0.001 A and 1 μA = 0.001 mA.

The unit of charge, the *coulomb* (C), is defined as the charge that passes along a wire in 1 second when the current is 1 ampere. Therefore, for a wire which carries a constant current,

the charge passing in a certain interval of time =
current in amperes × time interval in seconds

For example, the charge passing along a wire

- for a current of 2 amperes in 5 seconds is 10 coulombs,
- for a current of 2 amperes in 20 seconds is 40 coulombs,
- for a current of 20 amperes in 5 seconds is 100 coulombs.

Each electron carries a negative charge so small that 6.25 million million million are needed to make up 1 coulomb of charge. 6250 million million electrons pass through a 1 mA torch bulb every second when it lights normally.

Batteries and cells

In any circuit, electrons transfer energy from the battery to the components in the circuit. Each electron in an electric circuit gains energy as it passes through the battery, leaves the battery via the negative terminal, loses this energy as it passes round the circuit and re-enters the battery via its positive terminal. Each electron leaves the negative terminal of the battery with electrical potential energy which is used up and converted to other forms of energy as the electron passes round the circuit. For example, when an electron passes through a torch bulb, energy is transferred from the electron to the torch bulb. The effect of all the electrons passing through the torch bulb filament is to make it so hot that it emits light. The connecting wires are good conductors so no energy is given up in these wires, provided the current is not excessive.

A **battery** consists of two or more identical cells, each cell consisting of two electrodes in a conducting paste or liquid, the electrolyte, which reacts with the electrodes. The two electrodes are made from different materials such as graphite and lead, chosen because they react differently with the electrolyte. One electrode loses electrons to the electrolyte and the other one gains electrons from the electrolyte. Thus one electrode becomes positively charged and the other one becomes negatively charged. In use, a cell gradually deteriorates because the chemical reactions between the electrodes and the electrolyte convert the electrolyte and the electrodes into other substances which do not react. In a rechargeable cell, this chemical process can be reversed by connecting the cell to a battery charger which forces electrons into the cell at the negative electrode and out at the positive electrode. Disposable batteries and cells can not be recharged and are discarded when exhausted.

The **voltage** of a battery or a cell is the power in watts the cell can deliver to the rest of the circuit for every ampere of current passing through it. Thus a 12 volt battery connected in a circuit delivers 12 watts of power for every ampere of current passing through it. If it is connected to a suitable 3 A light bulb, the battery supplies 36 watts of power to the light bulb because the current is 3 amperes.

Voltage (in volts) = power delivered (in watts) per ampere of current

The capacity of a battery or cell is usually expressed in ampere hours. This is the number of hours for which it would be able to supply 1 ampere of current before it is exhausted. Thus a car battery with a capacity of 60 ampere hours would be able to supply a current of 1 ampere for 60 hours. Such a battery could supply a current of 2 amperes for 30 hours or a current of 5 amperes for 12 hours.

Cold start

One reason why cars sometimes fail to start in winter is because a bigger current than normal is needed to start a very cold engine. The oil in the engine becomes more sluggish as its temperature falls and so the moving parts are more difficult to move the colder the engine is. If the battery is not fully charged, it fails to start the engine because it cannot supply enough current. Switching on the car lights and the heater before starting the car would make starting even more difficult.

More about voltage

In an electric circuit which contains a battery, current passes round the circuit and through the battery, transferring energy at a steady rate from the battery to the rest of the circuit. The energy supplied every second by the battery is delivered to the rest of the circuit. The battery forces electrons round the circuit, delivering energy from the battery to the rest of the circuit. A battery in an electric circuit acts like a pump in a central heating system; the pressure of the pump forces water through one radiator after another round the system. There is a pressure drop across each radiator because the pressure at the inlet to each radiator is higher than at the outlet. The sum of the pressure drops round the system is therefore equal to the pump pressure. The pressure drop across each radiator is a measure of the work done by the pump to force the water through the radiator. Thus the sum of the pressure drops round the system is a measure of the total work done by the pump which is measured by the pump pressure. In the same way, the battery voltage is a measure of the power supplied by the battery which is equal to the power delivered to each part of the circuit. Hence the battery voltage is equal to the sum of the voltages round the circuit.

The voltage between any two points in an electric circuit is the power delivered per ampere of current to that part of the circuit between the two points.

For example, if the voltage between two points in a circuit was 20 volts, then every ampere of current passing from one point to the other would deliver 20 watts of power. Thus a current of 5 amperes passing from one point to the other would cause 100 watts (= 20 volts × 5 amperes) to be delivered to that part of the circuit.

Summary

Charge (in coulombs) = current (in amperes) × time (in seconds)

$$\text{Voltage or potential difference (in volts)} = \frac{\text{power delivered in watts}}{\text{current in amperes}}$$

Questions

Q1. Select the correct words from the list below to complete the passage after the list.

negative   positive   gains   loses

A polythene rod becomes negatively charged when it is rubbed with a dry cloth. This is because electrons carry a ________ charge and the rod ________ electrons and the cloth ________ electrons when they are rubbed together.

Q2. (a) State the unit of electric charge.
(b) How much electric charge passes through a torch bulb when the current is 0.030 A for
(i) 10 seconds, (ii) 5 minutes?

Q3. A 12 volt car battery is connected to a 12 V 3 A light bulb. The car battery has a capacity of 36 ampere hours. Calculate
(a) how long the battery can continue to supply this amount of current before it becomes flat,
(b) how much power the battery supplies to the light bulb.

Electric circuits

Circuit rules

A **series circuit** is one in which the same current passes through all the components in the circuit. In other words, all the electrons moving round such a circuit pass through every component in the circuit. The current may be measured using an *ammeter* which is a meter designed to measure current. The ammeter must be connected in series with the other components in the circuit so that the same current passes through all the components in the circuit.

A circuit with *components in parallel* with each other is one in which *part* of the current from the battery passes through each of the parallel components. Components in parallel in a circuit have the same voltage between the two points where they are connected to the circuit. Each electron moving round the circuit passes through one component or the other component when it moves from one point to the other. The voltage between two points in a circuit may be measured using a *voltmeter* which is a meter designed to measure voltage. To measure the voltage across a component, the voltmeter must be connected in parallel with the component.

- For components in series, the current is the same.
- For components in parallel, the voltage is the same.

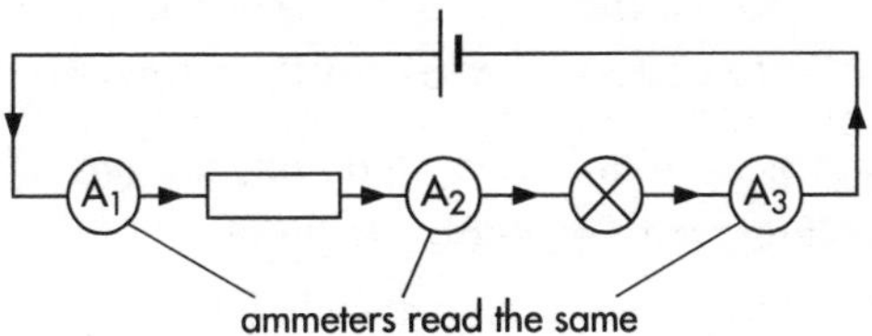

(a) Components in series

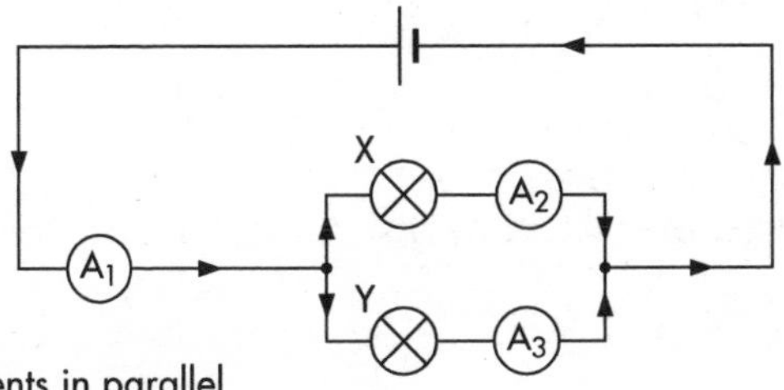

(b) Components in parallel
Figure 7C

Resistance

A potential difference or voltage exists between two points in an electric circuit if an electron has to do work to move from one point to the other. Conductors in a circuit resist the flow of electrons because the electrons repeatedly collide with the atoms of the conductor and lose energy in these collisions. The atoms of the conductor vibrate and the vibrations increase when the atoms gain energy from the electrons which collide with them. As a result, the conductor gains energy which it loses to the surroundings through heat transfer. In effect, the conductor resists the flow of electrons through it and the electrons must use some of their energy to pass through the conductor. Thus if a voltage exists between two points in any circuit, the electrons must use some or all of their energy to overcome the resistance of that part of the circuit to their passage from one point to the other.

The **resistance** of a circuit component is defined as

$$\frac{\textbf{the voltage across the component}}{\textbf{the current through the component}}$$

The unit of resistance is the *ohm* (symbol Ω) which is the amount of resistance between two points in a circuit when the voltage between the two points is 1 volt and the current is 1 ampere.

$$\textbf{Resistance (in ohms)} = \frac{\textbf{voltage(in volts)}}{\textbf{current (in amperes)}}$$

Note: the resistance formula above can be rearranged as voltage = current $\times$ resistance, or current = $\frac{\textit{voltage.}}{\textit{resistance}}$

Worked example

Calculate the resistance of a wire when the current through it is 2.0 A and the voltage across its ends is 3.0 volts.

Solution

$$\text{Resistance} = \frac{\text{voltage}}{\text{current}} = \frac{3.0 \text{ volts}}{2.0 \text{ amperes}} = 1.5\ \Omega$$

Circuit diagrams

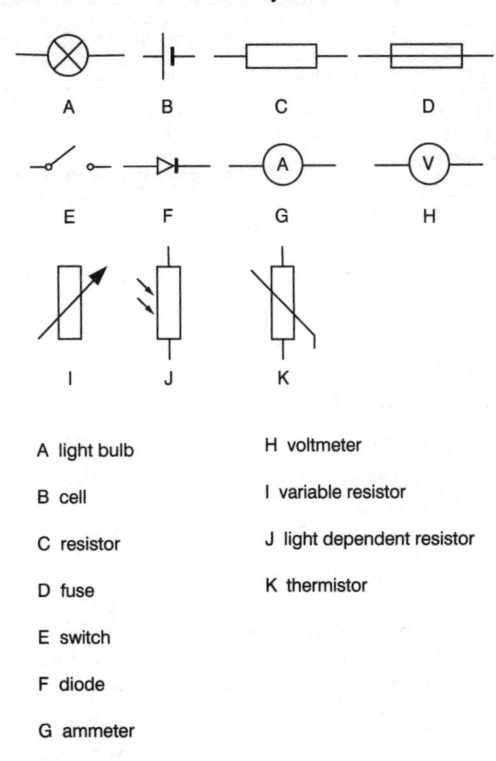

Figure 7D Electrical symbols

Standard symbols are used to represent components in circuit diagrams. The function of each component represented in Fig. 7D is described below if it has not been previously described.

- A *resistor* is a component designed to have a known resistance. For example, to make a 5.0 Ω wire-wound resistor from wire of resistance 2.5 ohms per metre, a length of exactly 2.0 metres would need to be used. The current through a resistor is constant for a fixed voltage across the resistor. The bigger the resistance of the resistor, the smaller the current is.

- A *variable resistor* is a component designed to change the current without necessarily changing the voltage. A variable resistor can be made by changing the length of a resistance wire in a circuit. If the length is increased, the resistance of the wire is increased and so the current becomes smaller.
- A *diode* allows current through in one direction only. A diode conducts when it is connected into the circuit in its forward direction as its resistance in the forward direction is very low. If the diode is then reversed, it does not conduct because its resistance in the reverse direction is very high.
- A *light emitting diode* (LED) emits light when current passes through it. LEDs are used as indicators in electronic circuits.

The electrical power equation

Because the voltage between any two points in a circuit is the power delivered per ampere, it follows that the power delivered for a certain current can be calculated by multiplying the voltage by the current. In other words,

$$\underset{\text{(in watts)}}{\textit{Power delivered}} = \underset{\text{(in volts)}}{\textit{voltage}} \times \underset{\text{(in amperes)}}{\textit{current}}$$

For example, an electric kettle that operates at a voltage of 230 volts and a current of 10 amperes would take 2300 watts of power (= 230 volts × 10 amperes) from the electricity supply when it is switched on. A car's electrical heater designed to operate at 12 volts and 5 amperes would take 60 watts of power (= 12 volts × 5 amperes) from the car battery if the engine was not switched on.

Summary

$$\underset{\text{(in ohms)}}{\text{Resistance}} = \frac{\text{voltage (in volts)}}{\text{current (in amperes)}}$$

$$\underset{\text{(in watts)}}{\text{Power delivered}} = \underset{\text{(in volts)}}{\text{voltage}} \times \underset{\text{(in amperes)}}{\text{current}}$$

Questions

Q4. A current of 2.5 A is passed through a 6.0 Ω resistor. Calculate (a) the p.d. across the resistor, (b) the power supplied to the resistor.

Q5. (a) Draw a circuit diagram to show a diode connected in series with a 1.5 V cell and a lit torch bulb.
(b) If the 1.5 V cell was connected in the circuit in the reverse direction, the torch bulb would not light. Why?

Q6. A set of Christmas tree lights contains 20 light bulbs in series, each rated at 12 volts and 6 watts.
(a) Calculate (i) the current through the light bulbs, (ii) the resistance of a single light bulb.
(b) Explain why none of the lights will light up if any one of the light bulbs fails.

Electricity at work

A magnetic compass near a wire is deflected when the current in the wire is switched on. This effect is because the current creates a magnetic field round the wire. With no current in the wire, the needle of the magnetic compass points north because of the Earth's magnetic field. If the compass is moved in the direction its needle points, it would follow a straight line leading northwards. Such a line is referred to as a magnetic field line, sometimes also called a magnetic line of force. Its direction is defined as the direction which a compass needle points. With the current on, the compass would need to move round the wire because the magnetic field lines are circles centred on the wire as shown in Fig 7E. If the current is reversed in the wire, the direction of the magnetic field lines round the wire reverses.

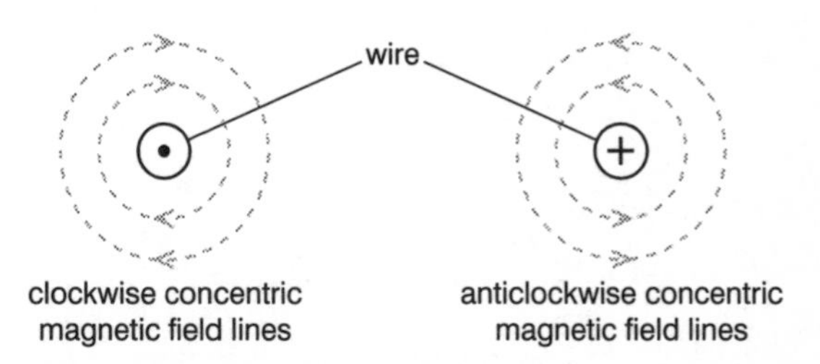

Figure 7E The magnetic field around a current carrying wire

Electromagnets

An electromagnet consists of a coil of insulated wire wound round an iron bar. When a current is passed through the coil, the iron bar is magnetized by the magnetic field due to the current. As a result, the iron bar is able to attract iron and steel objects held near its ends. When the current is switched off, the iron bar loses its magnetism. Electromagnets have many uses, from powerful electromagnets used to pick up and move cars in scrapyards to very sensitive electromagnets used to write data onto computer discs. Two further uses of electromagnets are described below.

- The relay: when current is passed through the coil of a relay, the electromagnet attracts an iron armature. The movement of this armature opens or closes a switch which is part of a different circuit. When the current is switched off, the armature springs back to its normal position and the switch reverts back to its original state.

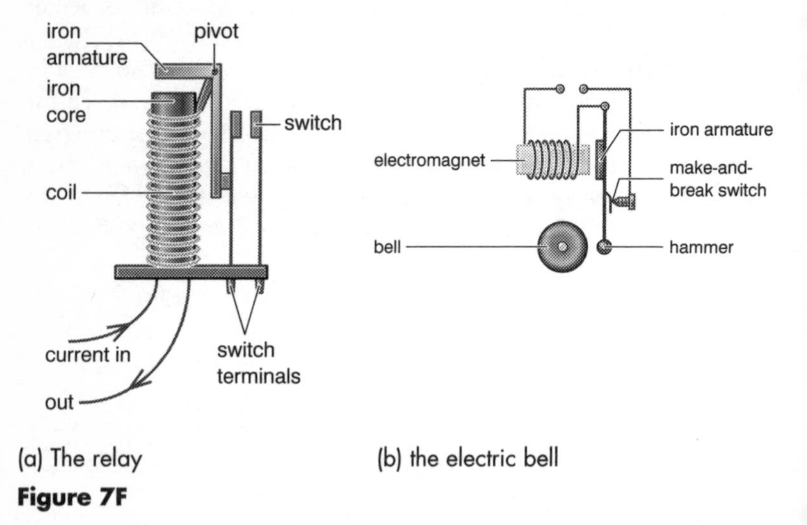

(a) The relay (b) the electric bell

Figure 7F

- The electric bell: the electromagnet coil is part of a 'make and break' switch. When current passes through the coil, the electromagnet attracts the iron armature which makes the hammer hit the bell. The movement of the armature opens the 'make and break' switch which switches the electromagnet off, allowing the armature to spring back and close the switch. Current then passes through the electromagnet and the sequence is repeated.

Electric motors

A wire near a magnet experiences a force when a current is passed along the wire. This effect is known as the *motor effect*. The reason for the effect is that the electrons passing along the wire are pushed sideways by the magnet, exerting a sideways force on the wire as a result. The effect depends on the angle between the wire and the lines of the magnetic field. For maximum effect, the wire needs to be at right angles to the lines of the magnetic field.

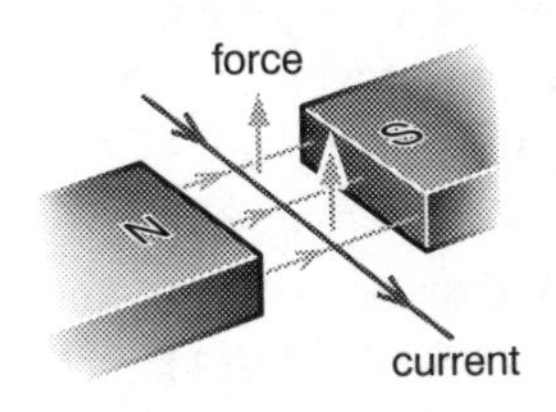

(a) The motor effect

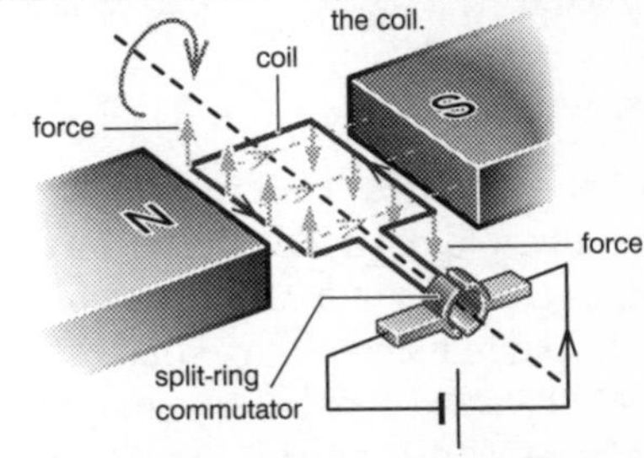

(b) The electric motor

Figure 7G

An electric motor consists of a coil of insulated wire which spins between opposite poles of a U-shaped magnet or electromagnet. When current is passed round the coil, the coil turns because the wires along one edge of the coil are forced up and the wires along the opposite edge are forced down. After half a turn, the current round the coil must be reversed otherwise the coil is turned back by the magnetic field. This reversal is achieved automatically by the action of the split-ring commutator which rotates with the coil. Conducting brushes made of graphite connected to the battery press against the commutator. Graphite is used as it is a conductor and provides contact with little friction.

The split-ring commutator has two functions:

1 It provides continuous contact between the battery and the coil as the coil turns,
2 It reverses the current direction round the coil every half turn so the coil is forced to spin in the same direction every half turn. Without this reversal of the current direction every half turn, the coil would reverse its direction repeatedly.

If the battery is reversed in the circuit, the motor spins in the opposite direction. The greater the current, the faster the rate of rotation of the coil. In a mains electric motor,

- an electromagnet is used instead of a permanent magnet,
- a number of coils at equal angles are wound on an iron core, thus providing smoother rotation. Each coil is connected to opposite sections of a multi-segment commutator.

Generators

Electricity can be generated in a coil by moving the coil in a magnetic field. Provided the coil wires cut the lines of the magnetic field, a voltage is induced in the coil. If the coil is part of a complete circuit, the induced voltage makes a current pass round the circuit. A cycle *dynamo* consists of a magnet which is forced to rotate near the end of a coil, causing a voltage to be induced in the coil. The faster the magnet rotates, the greater the voltage.

A voltage is induced in any wire in a magnetic field provided the wire and magnet are moving relative to each other and the lines of the magnetic field cut across the wire. This effect was discovered by Michael Faraday in 1831. The cause of the effect is that the electrons in the wire are made to move across the magnetic field as the wire and magnet move relative to each other. As a result, the magnetic field pushes the electrons along the wire. This push is sometimes referred to as an electromotive force and it only exists as long as the wire and magnet move relative to each other.

The *alternating current generator* consists of a rectangular coil of insulated wire that is forced to rotate between the poles of a U-shaped magnet. The wires along opposite edges of the coil cut across the magnetic field lines as the coil rotates, causing a voltage to be induced in the coil. The coil wires are connected to two slip rings on the axle of the coil. A graphite 'brush' pressed against each slip ring provides a continuous contact to an external circuit.

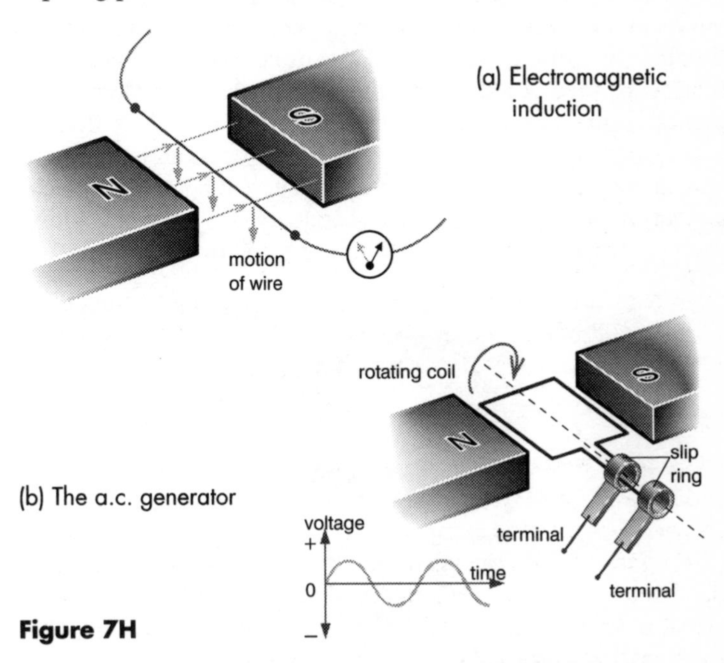

Figure 7H

The voltage reverses in polarity as each edge of the coil crosses the middle from one magnetic pole to the other pole every half turn. The voltage peaks when the coil edges are nearest the poles. Thus the voltage varies as shown, acting in each direction every half turn. This type of voltage is referred to as an *alternating voltage*. If the generator is connected to a suitable resistor, an alternating current is forced round the circuit, reversing its direction repeatedly. For each complete turn of the coil, the current reverses and reverses back again in a full cycle.

- The **frequency** of an alternating current or voltage is the number of cycles that take place each second. The unit of frequency is the hertz (Hz), where 1 hertz is equal to 1 cycle per second.
- The **peak voltage** of an alternating current or voltage is the maximum voltage or current in either direction.

Transformers

A transformer is designed to step an alternating voltage up or down. It consists of two coils of insulated wire, referred to as the primary and secondary coils, wound round the same iron core. When an alternating voltage is applied to the primary coil, an alternating voltage is induced in the secondary coil. The reason is that an alternating current passes through the primary coil and creates an alternating magnetic field in the core. This alternating magnetic field induces an alternating voltage in the secondary coil just as a rotating magnet in a dynamo induces a voltage in the coil of the dynamo.

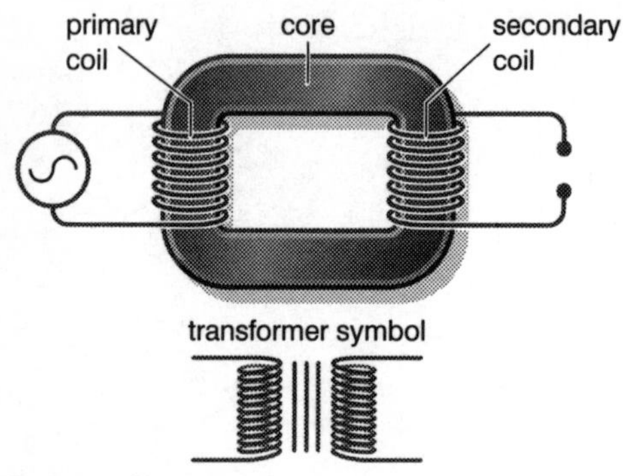

Figure 71 The transformer

The ratio of the secondary voltage to the primary voltage is equal to the number of windings or 'turns' on the secondary coil to the number on the primary coil.

$$\frac{\text{Secondary voltage}}{\text{Primary voltage}} = \frac{\text{number of secondary turns}}{\text{number of primary turns}}$$

- A step-up transformer has more windings on the secondary coil than on the primary coil. For example, if there are 20 times as many secondary turns as primary turns, the secondary voltage is 20 times the primary voltage.
- A step-down transformer has fewer windings on the secondary coil than on the primary coil. For example, if there are 20 times fewer secondary turns as primary turns, the secondary voltage is a twentieth of the primary voltage.

The Electricity Grid

This consists of a network of cables and transformers that connect power station generators to mains electricity users. Power is wasted in a cable if the current is too large because of the resistance of the cable. Because electrical power = current × voltage, then the same amount of electric power can be delivered using low current and high voltage as using high current and low voltage. Power station generators are connected to the grid system by means of step-up transformers. The high voltage on the cables means that the current through the cables is too low to heat the cables so little power is wasted. Step-down transformers at local sub-stations are used to reduce the voltage to a suitable level to match the needs of the user. Electricity supplied to homes and offices is at 230 volts.

Summary

The motor effect: a wire near a magnet experiences a force when a current is passed along the wire.

The transformer rule $\frac{\text{Secondary voltage}}{\text{Primary voltage}} = \frac{\text{number of secondary turns}}{\text{number of primary turns}}$

Questions

Q7. (a) State two uses of an electromagnet.
(b) With the aid of a diagram, explain the operation of an electric bell.

Q8. (a) Draw a labelled diagram of a simple electric motor and use your diagram to explain how the electric motor works.
(b) State and explain what happens if the battery connected to

a simple electric motor is (i) reversed, (ii) replaced by an alternating voltage supply.

(c) A mains electric motor has an electromagnet instead of a permanent magnet. The electromagnet and the armature coil are both connected to the mains which supplies alternating voltage. Explain why this type of motor spins in one direction only regardless of the direction of the current through the coil.

Q9. (a) With the aid of a diagram, describe the operation of an alternating current generator.

(b) A power station generator supplies electricity at 11 000 volts via a step-up transformer to the grid system at 132 000 volts.

(i) Calculate the turns ratio of the transformer.

(ii) The generator supplies 400 kilowatts of power to the grid. Calculate the current (1) through the generator circuit, (2) through the secondary coil of the transformer.

(iii) Explain why less power is wasted in the grid system by transmitting electricity at high voltage rather than low voltage.

Electricity in the home

Mains electricity is supplied from a local sub-station via a main cable, usually underground. The cable consists of two low-resistance insulated conductors. When an mains appliance is switched on, the appliance becomes part of a complete circuit consisting of the two conductors and the secondary coil of the sub-station transformer. One of the two conductors in the main cable, the *neutral* wire, is connected to the ground (i.e. 'earthed') at the sub-station so its voltage is zero. The other conductor, the *live* wire, alternates in voltage between about +340 volts and –340 volts.

The voltage of the mains is usually measured and stated in terms of the value of the steady (i.e. direct) voltage which would deliver the same power on average to any mains appliance. This 'equivalent' steady voltage works out at 230 volts. Thus a 230 volt 1000 watt mains electric heater would provide the same heating effect if connected to a 230 volt direct voltage supply as it would if connected to 230 volt mains.

Safety first

The live wires and any terminals connected to it are dangerous. Anyone who touches a live wire would suffer a fatal electric shock because the human body conducts electricity. No more than about 0.02 amperes through the body would cause a severe electric shock. The resistance of the body is about 1000 ohms. Voltages in excess of about 20 volts (= 0.02 amperes × 1000 ohms) are therefore dangerous. The live wire reaches about 340 volts every half cycle.

Domestic circuits

The main cable is connected to the circuits in a house via the distribution board. A main switch can be used to cut the electricity supply off at the distribution board. A fuse in each circuit at the distribution board cuts off the current to any circuit in the event of a fault. Each circuit in a house consists of insulated live and neutral wires connected to the live and neutral terminals of the distribution board.

- Lighting circuits in the house usually consist of low resistance cables. Individual sockets connected to the same cable are connected in parallel with one another.
- The ring main circuit in a house supplies electricity to the wall sockets of the house via insulated live and neutral wires. The ring main cable also includes a third insulated wire, the earth wire, which connects the metal chassis of any electrical appliance connected to the ring main to earth at the house. This is to prevent the metal chassis from becoming live if a live wire becomes loose in the appliance.

  The sockets of the ring main are connected in parallel with each other. Each plug used to connect an appliance to the ring main is fitted with its own fuse. The switch in each socket is always connected on the 'live side' of the socket. More current can be supplied via the ring main than via lighting circuits because the wires are thicker and therefore conduct better. Also, the ring main provides two routes for current from the distribution board for any appliance.

Each circuit from the fuse board is protected with its own fuse. If the fuse 'blows', the live wire is therefore cut off from appliances supplied by that circuit.

The mains cable from the substation to a building is connected via the electricity meter to the circuits in the building at the distribution fuse board. The live wire from the substation is connected via a main fuse to the electricity meter.

The two wires used to supply an electric current to an appliance are referred to as the **live** and the **neutral** wires. The neutral wire is earthed at the nearest mains substation.

Mains wires need to have as low a resistance as possible, otherwise heat is produced in them by the current. This is why mains wires are made from copper. All mains wires and fittings are insulated.

The fuse in a lighting circuit is in the fuse box. Each light bulb is turned on or off by its own switch. When the switch is in the off position, the appliance is not connected to the live wire of the mains supply.

A **ring main** is used to supply electricity to appliances via wall sockets. A ring main circuit consists of a live wire, a neutral wire and the **earth wire** which is earthed at the fuse board. The wires of a ring main are thicker than the wires of a lighting circuit because appliances connected to a ring main require more current than light bulbs do.

Each appliance is connected to the ring main by means of a three-pin plug which carries a fuse. An appliance with a metal chassis is earthed via the three-pin plug and the earth wire. This prevents the metal chassis from becoming live if a fault develops in the appliance. Appliances connected to the ring main can be switched on or off independently since they are in parallel with each other.

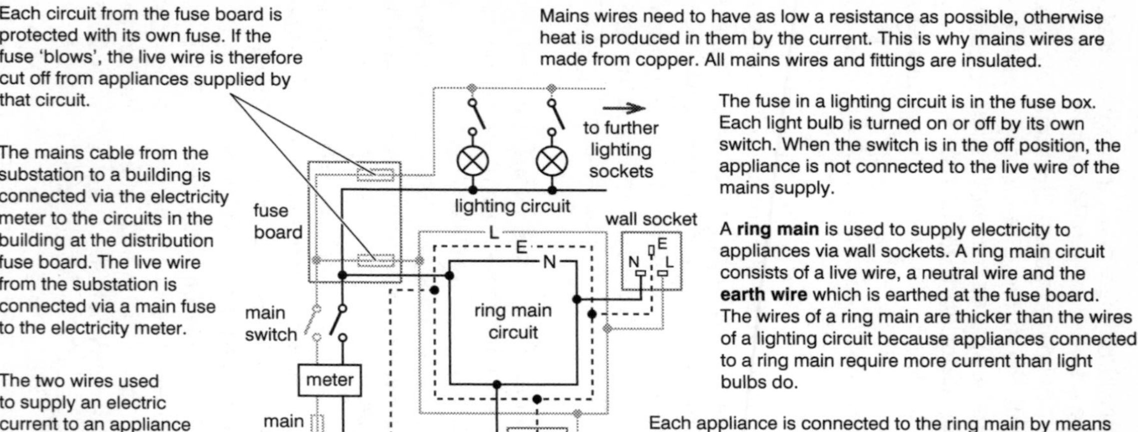

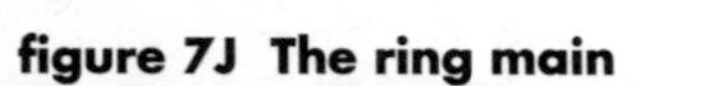

figure 7J The ring main

Fuses

A fuse is a thin wire designed to melt if the current through it exceeds a certain value referred to as the *fuse rating*. If a fuse melts, no current can pass through it because there is then a gap in the circuit. Thus a fuse is an essential safety feature of an electric circuit, intended to cut the current off if the current becomes too large. For example, an electric kettle fitted with a 13 A fuse operates normally when switched on provided the current does not exceed 13 A. If a fault in the kettle or the connecting cable causes the current to exceed 13 A, the fuse melts and cuts the current off completely.

Safety first again!

Always use a fuse of the recommended rating for a given appliance. For example, if a 5 A fuse is recommended for a hairdryer, a 13 A fuse would be unsafe as it would not melt if current could exceed the maximum safe value of 5 A and a 3 A fuse would melt every time the hairdryer was switched on to full power.

Electricity costs

Electricity meters measure electricity in units of kilowatt hours, where 1 kilowatt hour is defined as the energy used by a 1 kilowatt (= 1000 watts) appliance in 1 hour. A 3 kilowatt electric heater would therefore use 3 units (i.e. 3 kilowatt hours) of electricity in one hour. A 100 watt light bulb would use 2.4 units of electricity if switched on for 24 hours.

An electricity bill which shows that 2500 units of electricity were used in a certain period means that the energy used was the same as a 1 kilowatt appliance would use in 2500 hours. An electricity bill also states the unit cost of electricity which is the cost per kilowatt hour. Thus the cost of 2500 units of electricity at 5 p per unit would be £125 (= 2500 × 5 p).

Worked example

The electricity usage in a household was monitored for one week. During that time, the electricity usage was as follows:

A 3 kilowatt electric kettle was used 20 times for 6 minutes each time (i.e. 2 hours total).
An 800 watt microwave cooker was used 6 times for 10 minutes each time.

A 500 watt television was used for a total of 10 hours.
Four 100 watt electric lights were used for a total of 40 hours.
A 2 kilowatt electric heater was used for a total of 20 hours.
a) Calculate the number of units of electricity used by each appliance above,
(b) Calculate the total cost of the electricity supplied above for a unit cost of 5 p per unit.

Solution

(a) Kettle = 3 kW × 2 hours = 6 units; Microwave cooker = 0.800 kW × 1 hour = 0.8 units; Television = 0.500 kW × 10 hours = 5.0 units; lighting = 4 × 0.100 kW × 40 hours = 16 units; Heater = 2 kW × 20 hours = 40 units.
(b) Total number of units used = 6 + 5 + 0.8 + 16 + 40 units = 67.8 units

$$\text{Cost} = 67.8 \times 5 \text{ p} = 339 \text{ p}.$$

Summary

A **fuse** is designed to melt and cut the electric current off if the current exceeds the fuse rating.

Electricity meters measure electricity in units of kilowatt hours, where 1 kilowatt hour is defined as the energy used by a 1 kilowatt (= 1000 watts) appliance in 1 hour.

Questions

Q10. A 1000 watt 230 volt electric heater is to be fitted with a fuse. Which one of the following fuses should be chosen; 3 A, 5 A, 13 A.

Q11. (a) State the purpose of the earth wire in a ring main circuit.
(b) Why are the wires used for ring main circuits thicker than the wires used in lighting circuits?
(c) A short circuit occurs if a fault develops and a live wire or terminal makes contact with an earthed or neutral wire. Explain why (i) there is a risk of fire in a short circuit, and (ii) why a correctly fitted fuse prevents such a fire risk.

Q12. An electricity meter reading was 28501 units on a certain day and 28642 units one week later. (a) Calculate the cost of the electricity used in this period, given each unit cost 5.5 p.
(b) In the period above, a 6 kilowatt electric water heater was used for a total of 16 hours. Calculate how many units were due to the heater and the percentage of the total cost due to the heater.

8 THE NATURE OF LIGHT

In this chapter, we will start by studying reflection and refraction of light and the formation of images. We then move on to consider two different theories put forward over 300 years ago about the nature of light, both of which can explain reflection and refraction. One theory said that light was made up of tiny particles. The other theory said that light consists of waves. Which theory proved to be correct? We will look at the evidence for and against each theory and how more evidence was eventually obtained a century later which led to acceptance of one theory and rejection of the other.

The story of the nature of light had further twists in store with the discovery of photoelectricity at the end of the nineteenth century. Investigations on this effect produced observations which could not be explained until a new theory of light, namely the photon theory, was proposed by Albert Einstein, then a young physicist employed by the Swiss Patent Office. Einstein went on to develop his theories of relativity which include the astonishing outcome that nothing can travel faster than light. The speed of light in space at 300 000 kilometres per second is the cosmic speed limit. Even more astounding is Einstein's discovery that energy and mass are interchangeable on a scale according to his famous equation $E = mc^2$*, where c is the speed of light. The development of our understanding of light took place over several centuries and continues to develop as physicists strive to understand why light provides the link between energy and mass. So let's look at this story in a bit more detail.*

Properties of light

A laser beam or a beam of sunlight travels in a straight line. On a cloudy day, sunlight that breaks through a gap in the clouds can be seen as a straight beam extending to the ground. We talk about the Sun's rays but what do we mean by the word 'ray'? It seems like a very convenient word for the path that light takes when it radiates from a source (e.g. light from the Sun). So for convenience we will think of light in terms of rays. Also, let us assume for now that light travels in straight lines.

Reflection of light

Stand in front of a flat mirror and you will see an image of yourself. You will also be able to see beyond your own image the image of any other object behind you. Each image is the same distance behind the mirror as the object is in front. So if you stand 50 centimetres in front of a flat mirror, your image is 50 centimetres behind the mirror and is therefore 100 centimetres away from you. Step back from the mirror and your image steps back by the same distance.

Mirror images

Make a semi-transparent 'mirror' from cellophane film over an open box. Put the box on its side so the film is vertical and place a small object in front of it. You should be able to see an image of the object in the box. Move the object about until its image appears to be at the back of the box. You should find the distance from the object to the 'mirror' is then the same as the distance from the 'mirror' to the back of the box.

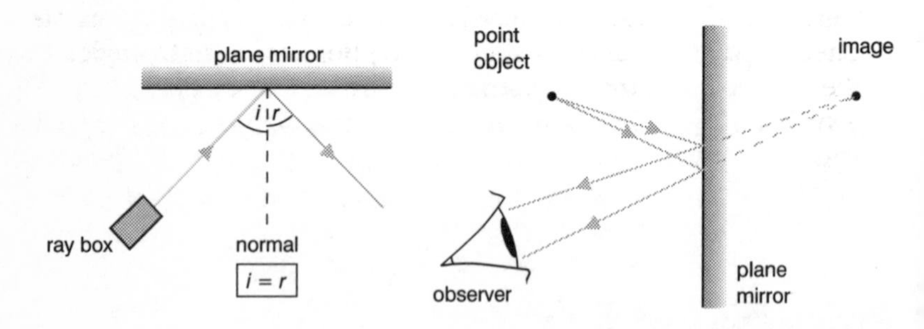

(a) The law of reflection (b) Image formation by a plane mirror

Figure 8A

When a ray of light is directed at a certain angle onto a flat mirror, the light ray reflects off the mirror at the same angle. In other words, the angle between the mirror and the light ray before reflection is the same as the angle between the mirror and the light ray after reflection. This statement is known as the *law of reflection* and it holds for any mirror. The law is usually expressed in the form

$$i = r$$

where i is the angle between the incident ray and the normal (which is the line at right angles to the mirror at the point of incidence) and r is the angle between the reflected ray and the normal.

Using this law, we can explain the distance rule by means of a ray diagram as shown in Fig. 8A. This diagram represents two light rays from a point object before and after reflection. The angle between each light ray and the mirror is the same before reflection as after reflection. Someone looking into the mirror along both reflected rays would see an image of the object at the position where the reflected rays appear to come from. The mirror in the ray diagram in Figure 8A is a line of symmetry. This is because each ray from the object to the mirror is at the same angle to the mirror as the corresponding 'apparent ray' from the image to the mirror. Therefore, the image and the object are the same distance from the mirror.

Refraction of light

Refraction of light is the change of direction of a light ray when it passes from one transparent substance to another transparent substance. For example, when light passes from air into a lens, its direction changes unless its initial direction was at right angles to the lens surface. Another example can be seen when the bottom of a swimming pool is viewed from above the water surface. The pool appears shallower than it really is. This is because light from a point on the bottom of the pool that passes through the surface non-normally is refracted away from the normal at the surface. As a result, the image of an object under water appears closer to the surface than the object really is.

Light is refracted

- towards the normal when it passes from air into a transparent substance.
- away from the normal when it passes from a transparent substance into air.

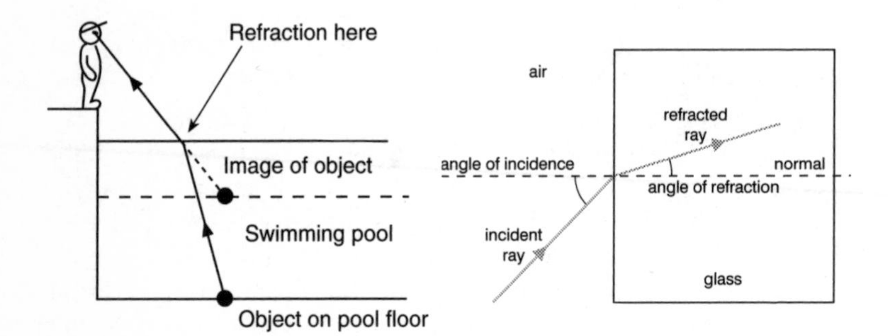

(a) Refraction at a water surface (b) Testing the law of refraction

Figure 8B

Refraction tests

1. Observe a drinking straw in a cup of water. Notice that it appears to bend at the surface. This is because light refracts at the surface.
2. Look at an object through a magnifying glass. You should see an enlarged image of the object because light from the object passing through the magnifying glass is refracted. The image is further from the lens than the object.
3. Place a glass of water on top of newspaper print and observe the print through the water. The print viewed through the water appears to be higher up than the rest of the print. Again, this is because light refracts at the surface.

The law of refraction was discovered by Snell in Holland in 1618. He measured the angle of refraction, r, between the normal and the refracted ray for different values of the angle of incidence, i (i.e. the angle between the normal and the incident ray). He found that the ratio $\sin i/\sin r$ is always the same for a given transparent substance. This ratio is referred to as the refractive index (symbol n) of the substance. Figure 8B (b) shows how to test the law of refraction.

$$\text{Refractive index, } n = \frac{\sin i}{\sin r}$$

Maths kit

For the right-angled triangle ABC shown in Fig. 8C, let the angle BAC be denoted by the greek letter θ (pronounced 'theta'). The following equations are used to define three trigonometry functions including sin θ (pronounced 'sine theta'):

$\sin\theta = o/h$,
$\cos\theta = a/h$,
$\tan\theta = o/a$,
where o is the length of the opposite side to θ, a is the length of the adjacent side to θ and h is the length of the side opposite the right angle.

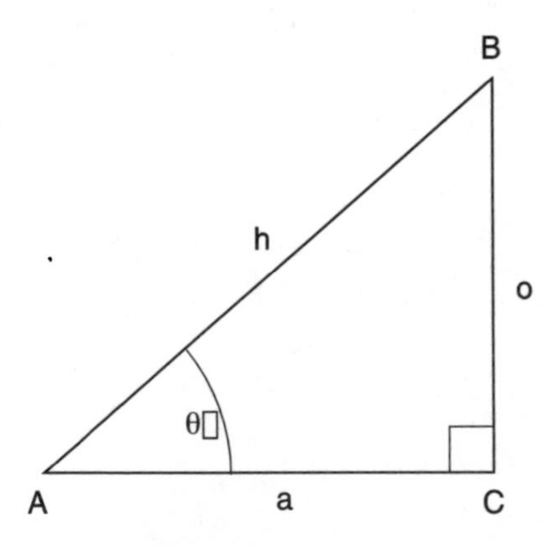

Figure 8C Trigonometry equations

Worked example

A light ray enters a glass block at an angle of incidence of 30°. The angle of refraction was found to be 22°. Calculate (a) the refractive index of the glass block, (b) the angle of refraction for an angle of incidence of 60°.

Solution

(a) $n = \sin 30 / \sin 22 = 0.50 / 0.34 = 1.5$

(b) Rearrange $n = \sin i/\sin r$ to give $\sin r = \sin i/n = \sin 60/1.5 = 35°$

Summary

Law of reflection: The angle between the mirror and the light ray before reflection is the same as the angle between the mirror and the light ray after reflection.

Snell's law of refraction: $\sin i/\sin r = n$, where i is the angle between the normal and the incident ray, r is the angle between the normal and the refracted ray, and n is the refractive index of the substance.

Questions

Q1. If you stand 0.60 m in front of a flat mirror, how far away from you is your own image?

Q2. State the law of reflection at a flat mirror.

Q3. Copy and complete the ray Fig. 8D showing the formation of an image by a flat mirror.

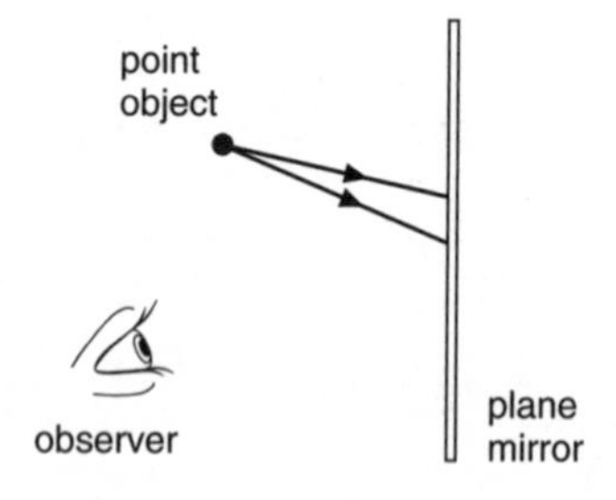

Figure 8D

Q4. (a) What is meant by refraction of light?

(b) A light ray is directed at a glass block of refractive index 1.5 at an angle of incidence of 60° as shown below. Calculate the angle of refraction of the light ray.

Q5. Copy and complete the path of the light ray through the glass block in the diagram.

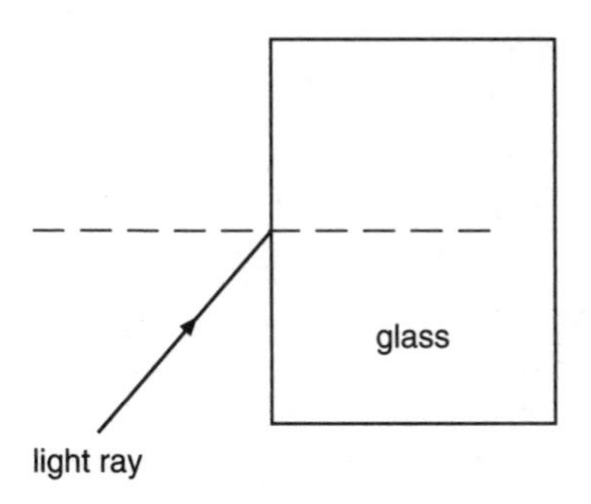

Figure 8E

Theories of light

Newton's theory of light

If you have played or observed a game of snooker you will know that a ball bounces off a wall at the same angle as it hits the wall – provided it is not spinning as it rolls towards the wall. In addition to establishing laws on motion and gravity, Sir Isaac Newton put forward a theory of light in which he proposed that light consists of tiny particles which he called *corpuscles*. Perhaps Newton had the idea in his mind about a ball bouncing off a wall for that was how he explained the reflection of light, supposing that the corpuscles of light bounced off a mirror like a ball bounces off a wall. Newton needed to assume the corpuscles did not lose any speed due to the impact so they moved away from the wall at the same angle as they had moved towards it.

Newton used his corpuscular theory to explain refraction of light as well as reflection. According to Newton, a corpuscle of light in air moving towards a transparent substance is attracted towards the substance. The force of attraction causes it to move faster in the substance than in air. More importantly, if its initial direction is not directly along the normal, the increase of speed causes its direction to change nearer to the normal.

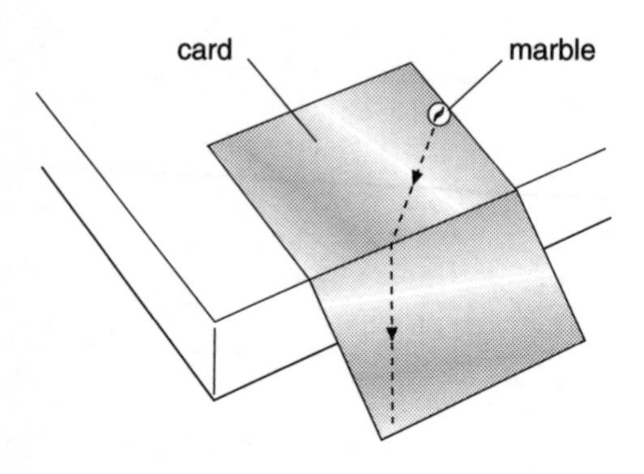

Figure 8F A refraction model

A model of refraction can be made using a marble and a piece of card folded down the middle. One half of the card needs to be fixed on a book with a slight slope so the other half of the card slopes steeply down to the table on which the book rests. Observe the progress of a marble rolled slowly across the card on the book towards the fold. When it passes over the fold onto the steeper section, its speed increases and its direction changes at the fold.

Huygens' wave theory of light

About the same time as Newton put forward the corpuscular theory of light, an alternative theory of light was put forward in Holland by Christiaan Huygens. According to Huygens, light consists of a wavemotion which moves through space or air or any transparent substance like waves move across a water surface. This imaginative theory also provided an explanation of the laws of reflection and refraction of light. Reflection of water waves is observed when sea waves bounce off a harbour wall. Refraction of water waves is the reason why waves on a beach usually run straight up the beach, regardless of the direction of the waves as they approach the beach. The waves slow down as the water becomes less deep and this reduction in speed causes them to move directly towards the beach. This situation is not unlike a front-wheel drive vehicle that veers off a road onto muddy ground. The vehicle changes direction as it goes off the road because one of the front wheels leaves the road and loses some of its grip before the other one does. To explain refraction of light, Huygens needed to assume that light travels slower in a transparent substance than in air. In comparison, Newton needed to assume light travels faster in a transparent substance than in air.

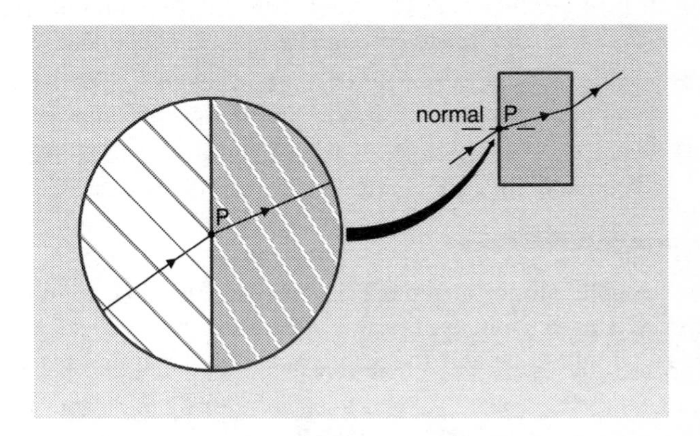

Figure 8G Using wave theory to explain refraction

Young's double slits experiment

Reflection and refraction of light can both be explained using either Newton's corpuscular theory or Huygens' wave theory of light. Newton's theory assumes that light travels faster in a transparent substance than in air whereas Huygens' theory assumes it travels slower in a transparent substance. Which theory is correct? When the theories were compared in the seventeenth century, most people believed Newton because the speed of light could not be measured and Newton's scientific reputation was much greater than that of

Huygens. Newton's theories of motion and gravity were outstandingly successful in explaining the motion of objects in all possible situations. So Newton's theory of light was accepted for over a century even though there was no direct evidence for the corpuscular theory of light.

The idea that light consists of tiny particles held sway until the first decade of the nineteenth century when Thomas Young at the Royal Institution in London used light to demonstrate the phenomenon of **interference**. Young showed that if a narrow source of light is observed through two closely spaced slits, a series of bright and dark bands (referred to as 'fringes') are seen. The fringes are replaced by a broad band of light if one of the two slits is blocked. Thus the dark bands are formed where light from one slit cancels out light from the other slit. The light from the two slits is said to interfere

- positively where a bright fringe is formed. The light from each slit reinforces the light from the other slit.
- negatively where a dark fringe is formed. The light from each slit cancels the light from the other slit.

Observing interference

Use the double slit arrangement shown in Fig. 8H to observe interference fringes using light from a torch lamp adapted as a narrow source. You ought to be able to see bright and dark parallel fringes.

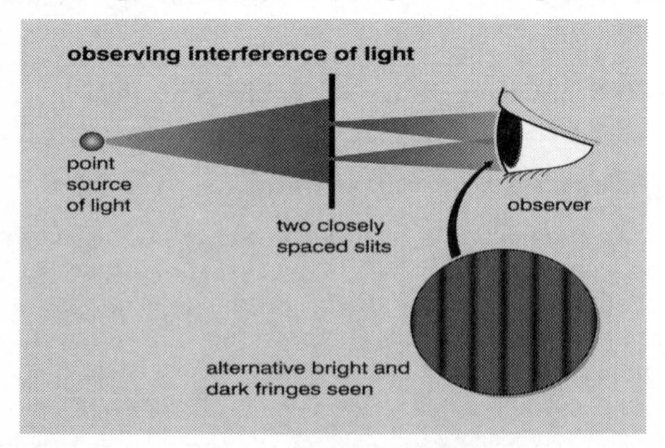

Figure 8H Observation of interference fringes

The phenomenon of interference can also be demonstrated by sending straight waves on a water surface towards two narrow closely spaced gaps in a barrier in the water. The waves that pass through the gaps spread out and overlap. In the overlap region, wave crests and troughs from one slit overlap and pass through crests and troughs from the other slit.

- Reinforcement occurs where a crest meets a crest or a trough meets a trough. With light, this occurs at each bright fringe.
- Cancellation occurs where a crest meets a trough. With light, this occurs at each dark fringe.

Young demonstrated the phenomenon of interference of light before an audience of invited guests at the Royal Institution. However, he was unable to convince them that light was wave like in its nature as they preferred to believe that some unknown property of corpuscles would explain the phenonemon of interference. The dispute about whether light consists of waves or particles was only settled several decades later when it was shown clearly that light travels slower in water than in air – as predicted by Huygens over a century earlier!

The wavelength of light, the distance from one crest of a wave to the next crest, depends on its colour. By making measurements using Young's slits arrangement, the wavelength of each colour can be determined. The wavelength of each colour is shown in Fig. 8J. Note that the wavelength decreases from about 0.0007 millimetres for red light to about 0.0004 millimetres from violet light. Two thousand wavelengths of yellow light would fit into the space between adjacent millimetre marks on a ruler.

The smallness of the wavelength of light is the reason why wave effects are difficult to demonstrate with light.

The speed of light

Light travels through space at a speed of 300 000 kilometres per second. Light from the Sun takes about 500 seconds to reach the Earth. Light from the most distant galaxies reaching us now was emitted thousands of millions of years ago. The first accurate measurement of the speed of light was made by Fizeau in France in 1849. He observed light from a narrow source of light after it had travelled a distance of over 17 kilometres. A narrow beam of light from the source was directed at the teeth of a rotating cog wheel which chopped the beam into pulses. The beam was then directed to a distant mirror where it was reflected back to the edge of the cog wheel. The beam passes back through the edge of the cog wheel if a gap between the teeth is present when each pulse returns to the wheel. Figure 8I shows the idea. If the wheel is turned faster and faster from zero speed, an observer looking at the image of the light source along the returning beam would see the image at certain rotation speeds only. At each such rotation speed, the time taken by the beam to travel from the wheel and back again is equal to the time taken for a gap to be replaced by another gap at the light beam.

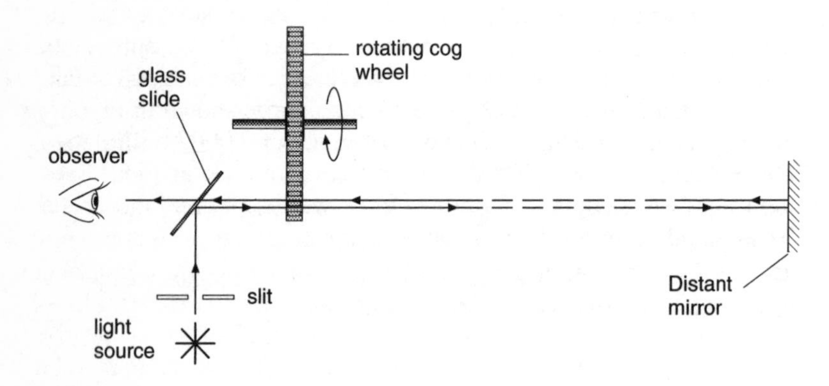

Figure 8I The speed of light

For a wheel with N teeth turning at a rotation frequency of f turns per second,

- the time taken for 1 rotation of the wheel = $1/f$
- the time taken for one gap to replace the next gap, t = time for 1 rotation / $N = 1/fN$

If the light travels a distance D in this time (from the wheel and back), the speed of light, c = distance / time taken = $D/(1/fN) = DfN$

Fizeau used a wheel with 720 teeth and found that the lowest frequency at which the image reappeared was 25.2 rotations per second when the total distance from the wheel and back was 17.3 km. Prove for yourself that these measurements give a value for the speed of light of 315000 kilometres per second. Fizeau used this method to show that light travels more slowly through water than through air. Improved methods for the measurement of the speed of light were subsequently devised. The speed of light in a vacuum is now defined as 299792.458 kilometres per second.

Electromagnetic waves

Water waves are disturbances that travel across the surface. Sound waves in air are pressure variations that travel through the air. Seismic waves are vibrations that travel through the Earth after being created in an earthquake. What are light waves? In the nineteenth century, it was generally thought that light consists of vibrations in an invisible substance referred to as 'ether' which was thought to fill space. In 1862, a mathematical theory of light waves was published by the Scottish physicist, James Maxwell. He showed that light consists of vibrating electric and magnetic fields in which the electric vibrations generate magnetic vibrations which generate electric vibrations, etc. Such waves were referred to as electromagnetic waves. Maxwell combined the theory of electric fields with the theory of magnetic fields to show that such electromagnetic waves ought to travel through space at a speed of 300000 kilometres per second. He knew that light travels at this speed through space so he concluded that

- light consists of electromagnetic waves, and
- electromagnetic waves exist beyond both ends of the visible spectrum.

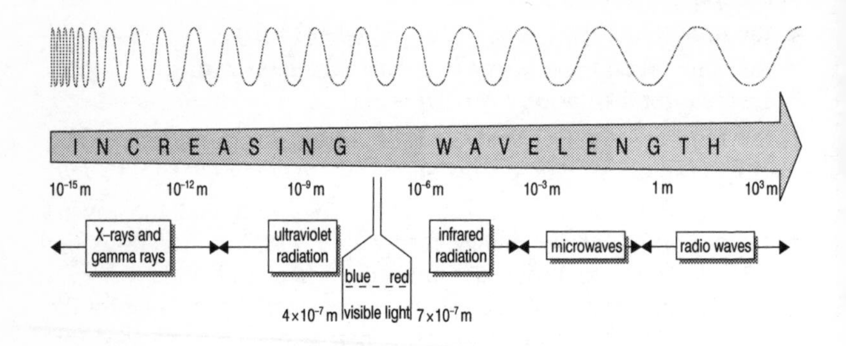

Figure 8J The electromagnetic spectrum

When Maxwell published his theory of electromagnetic waves, it was immediately realized that infrared radiation which had been detected beyond the red part of the visible spectrum consists of electromagnetic waves beyond the red end of the visible spectrum. In addition, ultraviolet radiation which was known to lie beyond the violet part of the visible spectrum must also consist of electromagnetic waves.

Could there be electromagnetic waves even longer in wavelength than infrared radiation? What about beyond ultraviolet radiation? Several decades after Maxwell published his theory, the German physicist Heinrich Hertz discovered how to produce and detect electromagnetic waves much longer in wavelength than infrared radiation. These waves became known as *radio waves* and within a few years they were being used to transmit wireless signals between Britain and America.

Tuning in

When you tune in to a radio or TV station, you are adjusting a receiver circuit so it will accept radio or TV waves of a certain wavelength only. The **wavelength** of a wave is the distance along the wave from one wavepeak to the next wavepeak. The Greek symbol λ (pronounced 'lambda') is used for wavelength. The tuner of a radio receiver usually displays the wavelength or the frequency of the waves which the tuner accepts.

- The frequency is the number of complete cycles of waves passing a point each second, where one cycle is from one wave crest to the next wave crest. The unit of frequency is the hertz (abbreviated as Hz), where 1 Hz = 1 cycle per second. Local radio stations broadcast at frequencies of about 100 megahertz (MHz), where 1 MHz = 1 million hertz.
- The speed of waves is the distance per second travelled by a wave crest. For waves moving at speed υ, a wave crest would travel a distance in 1 second equal to υ. The number of wavelengths in this distance is equal to υ / λ, where λ is the wavelength of the waves. Hence the frequency, f, of the waves, which is the number of waves passing a given point in 1 second, is equal to υ / λ.

Frequency, f = speed of waves, υ / wavelength λ

Hence for electromagnetic waves of wavelength λ, the frequency of the waves = c / λ, where c is the speed of electromagnetic waves. For transmission through air or through space, c = 300000 kilometres per second.

Worked example

Calculate (a) the frequency of radio waves of wavelength 100 metres in air, (b) the wavelength of radio waves in air of frequency 100 MHz. The speed of electromagnetic waves in air = 300000 kilometres per second.

Solution

(a) Frequency, f = speed, c / wavelength, λ = 300000 × 1000 metres per second / 100 metres = 3000000 Hz = 3.0 MHz.

(b) Multiplying both sides of $f = c / \lambda$ by λ gives $f\lambda = c$.
Dividing both sides of $f\lambda = c$ by f gives $\lambda = c / f$.
Hence λ = 300000 × 1000 metres per second / 100000000 Hz = 3.0 m.

Frequency bands

	frequency range	uses
long wave (LW)	up to 300 kHz	international AM radio
medium wave (MW)	300 kHz–3 MHz	AM radio
high frequency (HF)	3–30 MHz	AM radio
very high frequency (VHF)	30–300 MHz	FM radio
ultra high frequency (UHF)	300–3000 MHz	TV broadcasting, mobile phones
microwave	above 3000 MHz	satellite TV, global phone links
light	500 THz approx	fibre optic communication links

Note: 1 MHz = 1 000 000 Hz. 1 THz = 1 million MHz.

Figure 8K Frequency bands

X-rays

A few years after the discovery of radio waves, the German physicist Rontgen discovered how to produce and detect electromagnetic waves much shorter than ultraviolet radiation. These waves became known as *X-rays*. Their discovery received great publicity as newspapers claimed they could be used to 'see' through objects. In fact, X-ray machines were quickly put to use in hospitals to photograph broken bones in limbs. This is possible because X-rays are absorbed by dense materials such as bone and they pass through soft material such as human tissue. By directing a beam of X-rays at a photographic film in a light-proof wrapper, the film is blackened where X-rays reach it because the X-rays, unlike light, can pass through the wrapper. If a limb is placed in the path of the beam before it reaches the film, a shadow of the bone in the limb can be seen on the film when it is developed because the bone prevented X-rays from reaching the film.

In the same decade as X-rays were discovered, the discovery of radioactivity by Becquerel in France led to the conclusion that radioactive substances can emit electromagnetic waves even shorter than X-rays. This type of radiation is known as gamma radiation (or γ radiation using the Greek letter γ, pronounced 'gamma'). Maxwell's theory of electromagnetic waves was thus

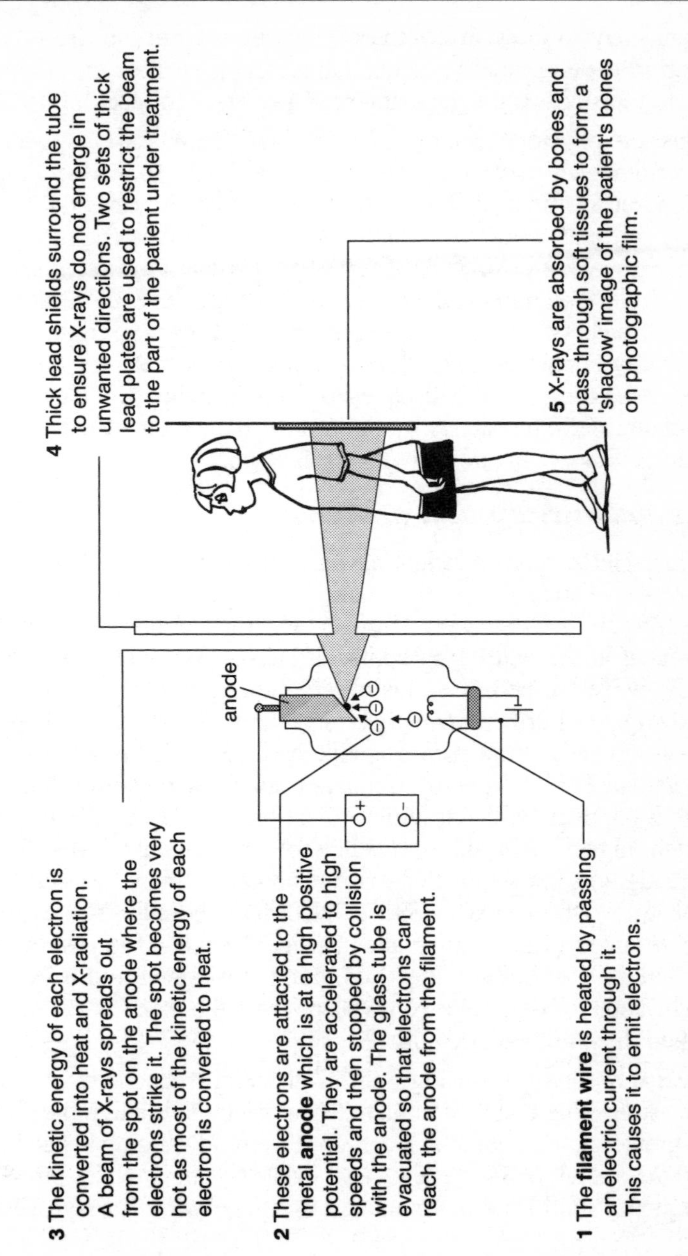

Figure 8L X-rays

confirmed by these discoveries. Predictions from the theory were used to develop many practical applications such as the design of radio transmitters on ships and of X-ray tubes for use in hospitals.

Classical physics based on Newton's laws and Maxwell's theory of electromagnetic waves was able to account for every known phenomenon or observation. Many physicists towards the end of the nineteenth century thought the laws of nature had more or less been discovered using the theories of classical physics. It seemed that little else remained to be explained although more accurate measurements of the properties of materials and light would probably enable them to justify their continued efforts. Although some niggling minor discoveries were proving troublesome to explain, many physicists by the turn of the century were satisfied that the laws of nature were mostly known.

Photoelectricity and photons

When Hertz was investigating how to produce and detect radio waves, he noticed that the sparks induced by radio waves in a 'spark-gap' detector were stronger when ultraviolet radiation was directed at the spark gap contacts. Hertz was more interested in radio waves so he passed his observation on to other physicists to investigate. Further investigations showed the effect happens because a metal emits tiny negative particles called *electrons* when illuminated with light of frequency above a certain value. The effect happens with ultraviolet radiation because its frequency is much higher than the frequency of any colour of light. The investigators knew that electrons are contained in every atom. They also knew that metals conduct electricity because some of the electrons in a metal move about inside the metal, not confined to individual atoms. Now they had found that light could be used to make some of these electrons escape from the metal. This effect is called the *photoelectric effect*.

Precise observations using light of different frequencies produced the astonishing conclusion that the frequency of the light needed to be greater than or equal to a certain 'threshold' value that depended on the metal being tested. The existence of such a threshold frequency could not be explained using the wave theory of light.

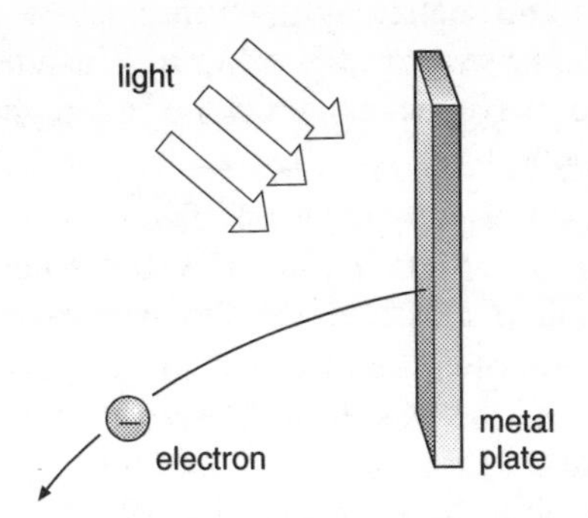

Figure 8M Photoelectricity

According to wave theory, the effect should happen regardless of the frequency of the light. It was expected that the effect would happen more slowly the lower the frequency of light used – but it would nevertheless happen according to wave theory. However, the experimental results showed that electrons were not emitted at all if the frequency of the light was below the threshold frequency. In 1905, this troublesome problem from the last few years of the nineteenth century was to shatter the well-established classical wave theory of light.

Einstein's photon theory of light

Einstein was an awkward student at the University in Zurich for he kept asking questions that his professors were unwilling or unable to answer. He graduated with a degree in physics in 1902 and eventually secured a post as a patent officer in Berne, Switzerland. He attended to his official duties efficiently and he was able to pursue his ideas about physics in his spare time. In 1905, he wrote several scientific papers which revolutionized physics and which recast the laws of physics. We shall return to his mind-boggling ideas about space and time in Chapter 10. For the moment, we will concentrate on his explanation of photoelectricity for which he was awarded the Nobel Prize for physics in 1921.

Einstein explained photoelectricity in his 1905 paper on the interaction of light and matter by inventing a new theory of light which he called the *photon theory of light*. Einstein assumed that light is composed of wavepackets which he called photons. The key points of photon theory are:

1. Each photon is a packet of electromagnetic waves moving in a particular direction, not spreading in all directions as in classical wave theory.
2. The energy of a photon is in proportion to the frequency of the waves in the wavepacket. Einstein used an earlier idea from Planck in Berlin about energy which we shall meet in Chapter 10 and assumed the energy E of a photon is given by the equation

$$E = hf,$$

where h is a constant referred to as the Planck constant.

To explain photoelectricity using photon theory, Einstein said that

1. an electron in a metal needs a minimum amount of energy to escape. This amount of energy is called the work function of the metal.
2. an electron near the surface can escape if it gains energy equal to or greater than the work function of the metal,
3. an electron is able to escape from the surface if it absorbs a single photon of energy greater than the work function of the metal.

Thus if an electron near the surface absorbs a photon of energy hf, it can escape if the photon energy hf is greater than or equal to W, the work function of the metal. The threshold frequency of the light therefore is equal to W / h, corresponding to $hf = W$.

Hence the frequency of the incident light, f must be greater than or equal to the threshold frequency for photoelectric emission to occur. Thus Einstein's photon theory provides an explanation of photoelectricity. Further predictions about the interaction of light and matter were confirmed by more investigations after 1905 and by 1921 the photon theory had been fully accepted.

The story of light so far

In the seventeenth century, Newton thought that light was corpuscular in nature. Huygens' wave theory of light was not accepted until the early decades of the nineteenth century when light was shown to travel slower in water than in air, as predicted by Huygens but not by Newton. Interference of light was discovered by Young in the first decade of the nineteenth century but was not accepted at that time as evidence for the wave theory of light. The particle nature of light was reintroduced as a result of the discovery of photoelectricity and its subsequent explanation by Einstein in terms of his photon theory. Light has a dual nature which is wave-like in interference experiments and particle-like in photoelectricity. This dual nature of light was to provide a path in a direction completely unexpected by physicists at the end of the nineteenth century. Within two decades, the classical laws of physics had given way as a revolution swept through physics and our perception of matter and energy. We will meet the new physics, known as quantum physics, in Chapter 10.

Summary

Equation for the frequency of a wave: frequency = speed of the waves / wavelength.

Electromagnetic waves

1. The spectrum of electromagnetic waves in order of increasing wavelength; gamma rays and X-rays; ultraviolet radiation; visible light; infrared light; microwaves; radio waves.
2. Visible light covers the wavelength range from about 0.0004 mm for violet light to about 0.0007 mm for infrared light.
3. All electromagnetic waves travel at a speed of 300 000 kilometres per second.

Photon theory

1 Light consists of photons which are wavepackets of electromagnetic radiation. The energy of a photon $E = hf$, where h is the Planck constant and f is the frequency of the light.

2 Photoelectricity is explained by the photon theory by assuming that the energy gained by a conduction electron in a metal which absorbs a photon ($= hf$) is sufficient to enable the electron to leave the interior of the metal if $hf >$ the work function of the metal.

Questions

Q6. (a) What was the nature of light according to Newton?

(b) Why was the wave theory of light eventually accepted?

Q7. Explain why a pattern of bright and dark fringes is observed when light from a narrow slit is observed through a pair of closely spaced double slits.

Q8. (a) What is a photon of light?

(b) What is the photoelectric effect?

(c) What aspect of the photoelectric effect could not be explained by wave theory?

(d) How does the photon theory explain the photoelectric effect?

Q9. List the following parts of the electromagnetic spectrum in order of increasing wavelength:

gamma rays, infrared radiation, microwaves, radio waves, ultraviolet radiation, visible light.

Q10. Light from a laser beam has a wavelength of 0.0006 mm.

(a) What is the colour of this light?

(b) Calculate the frequency of light of this wavelength. The speed of light is 300000 kilometres per second.

(c) What type of radiation is beyond the long wavelength end of the visible spectrum?

9 MATERIALS AND MOLECULES

Many new materials have been discovered or invented in the past century by scientists extending their knowledge about existing materials. Before the Scientific Age, most materials were natural products or made from minerals. The Scientific Age brought new processes, new products and new discoveries which have been put to use to make our lives better. A list of materials used by each of us every day would be very long. All the more surprising then that all materials and substances, manufactured or natural, are made from no more than 92 basic substances known as chemical elements. Before the Scientific Age, all substances were thought to be made from four basic elements, namely earth, water, air and fire. Progress to our present knowledge is due to the efforts of many scientists whose experiments and observations gave rise to scientific theories to explain the experimental data and then to further predictions and more experiments. New materials continue to be discovered in the same way, often leading to new devices and products. Understanding the properties of any substance requires underpinning knowledge and a sound grasp of certain key ideas. In this chapter, we will look at the development of these ideas and see how they provide the foundations of the science of materials.

Atoms and molecules

How would you describe the differences between solids, liquids and gases? Solids have their own shape whereas liquids and gases do not. Liquids and gases can flow whereas solids cannot. Liquids and solids have surfaces whereas gases do not. The three states of matter, solid, liquid and gas, are described as physical states because a substance can be changed from one state into another then back into its former state by heating and cooling. For example, ice melts and becomes water if heated. In contrast, chemical changes involve reactions between substances that cannot be reversed by simple physical means such as heating and cooling. For example, water and carbon dioxide are formed when a candle burns but candle wax cannot easily be formed from water and carbon dioxide. The obvious differences between the three states of matter are explained using two key ideas:

- all substances consist of molecules. A molecule is the smallest particle of a substance that can exist independently. The molecules of a pure substance are identical to one another and differ from the molecules of any other substance.
- molecules link together in the liquid state and the solid state because they exert forces on each other. These forces are referred to as *bonds* because they bind the substance in the liquid or the solid state. Energy supplied to a solid at its melting point causes it to melt because the molecules gain sufficient energy to break free from each other. Energy supplied to a liquid at its boiling point causes it to vaporize because the molecules gain sufficient energy to move away from each other.

Compounds and elements

A *compound* is a pure substance which consists of one type of molecule only. A compound can be broken down into other substances.

An *element* is a pure substance that cannot be broken down into other substances. There are just 92 naturally occuring elements. Further short-lived elements have been discovered in nuclear reactions but none of these is found naturally.

An *atom* is the smallest part of an element that is characteristic of the element. The atoms of an element are identical to one another and differ from the atoms of any other element. The lightest atom is the hydrogen atom. The heaviest naturally occuring atom is the uranium atom which is 238 times heavier than the hydrogen atom.

A *molecule* consists of two or more atoms joined together by force bonds. Each type of molecule consists of a fixed number and type of atoms. For example, every carbon dioxide molecule consists of one carbon atom and two oxygen atoms. Its chemical formula is therefore written as CO_2.

The idea of atoms was first put forward in Ancient Greece by the philosopher Democritus (470–400 BC) who considered objects to be made of tiny indivisible and indestructible particles he called atoms. His theory was rejected by Aristotle in favour of the theory of ' the elements', earth, water, air and fire. The atomic theory was established in its modern form in the early nineteenth century by John Dalton (1766–1844). He knew from the work of other scientists that the composition by weight of the elements in a given compound is always the same. He realized that this is because a compound consists of compound atoms or molecules, each molecule of a compound being composed of a fixed number of atoms of the elements which the compound can be broken down into. He worked out the relative weight of each type of atom, based on one unit of atomic weight for the hydrogen atom. He drew up a table of atomic weights for all the known elements and used it to explain why elements combine in simple proportions when they form compounds. Dalton's atomic theory led to the Periodic Table which is the basis of modern chemistry and which can only be explained using quantum mechanics (see p. 165).

How small is a molecule?

Molecules are much smaller than the tiniest grain visible with an optical microscope. You can see for yourself just how small a molecule is by making an oil film one molecule thick as follows:

1. Pour water into a tray on a draining board until the water brims over. When the water has settled, clean the water surface by moving a ruler across the surface from one end of the tray to the other end. Then sprinkle some very light dry powder on the water surface.
2. The next step is to dip a needle into some oil to obtain a tiny oil droplet on the end of the needle. The needle tip is then touched on the water at the centre of the tray. You should see a large patch of oil spread over the water, pushing the powder away. Provided the patch does not reach the sides of the tray, it is no more than one molecule thick.

Inside an atom

By the end of the nineteenth century, scientists knew that atoms are not indivisible or indestructible. Electrons were discovered as a result of experiments on electrical discharges through gases at very low pressures. These electrons are pulled out of the gas atoms by the strong electrical field applied to the discharge tube. Regardless of which type of gas was used, it was discovered that identical negatively charged particles were produced as well as positively charged atoms of differing weights. It was therefore concluded that these negatively charged particles, referred to as **electrons**, are in every type of atom. Further investigations on radioactivity (see Chapter 11) led to the conclusion that

- every atom contains a positively charged nucleus where most of its mass is concentrated.
- the nucleus is composed of two types of particles, **protons** and **neutrons**. The nucleus of the hydrogen atom is a single proton.
- the proton is a positively charged particle and is slightly lighter than the neutron which is uncharged.
- electrons move in the space round the nucleus at relatively large distances. The charge of the electron is equal and opposite to the charge of the proton. The proton is about 2000 times lighter than the electron.

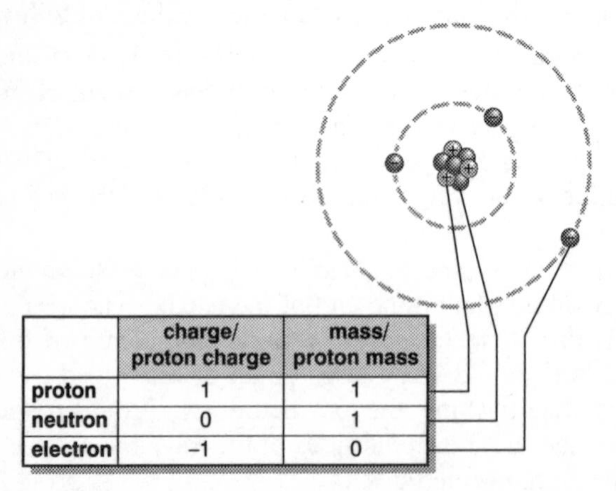

	charge/ proton charge	mass/ proton mass
proton	1	1
neutron	0	1
electron	–1	0

Figure 9A The structure of the atom

An atom is represented by the symbol $^{A}_{Z}X$ where Z is the proton number of the atom and A, the mass number, is the number of protons and neutrons in the nucleus. The mass of an atom is approximately equal to A times the mass of a hydrogen atom. This is because A is the number of proton and neutrons in its nucleus and each proton or neutron has a mass approximately equal to the mass of a hydrogen atom. For example, the symbol $^{7}_{3}Li$ represents an atom of lithium with three protons and four neutrons in its nucleus and three electrons moving about the nucleus. The mass of such an atom is approximately seven times the mass of a hydrogen atom.

Isotopes

All atoms of the same element contain the same number of protons in the nucleus of each atom. The number of neutrons in each atom of an element can differ. Atoms of an element with different numbers of neutrons are referred to as **isotopes**. For example, natural uranium mostly consists of the isotope $^{238}_{92}U$ and a small proportion of the isotope $^{235}_{92}U$. Both types of atoms are uranium atoms, each nucleus containing 92 protons. However, the isotope $^{238}_{92}U$ contains three more neutrons than the isotope $^{235}_{92}U$.

Note that the number of electrons in an uncharged atom is equal to the number of protons in the nucleus. The chemical properties of an element are the same for all the isotopes of the element. This is because chemical reactions are determined by the electrons in an atom. Atoms of the same element undergo the same chemical reactions because each atom has the same electron arrangement even if the atoms are different isotopes of the same element.

The atomic scale of mass

The mass of an atom or a molecule is usually expressed in atomic mass units (symbol u) where $1 \text{ u} = 1.66 \times 10^{-27}$ kilograms. One unit of atomic mass (1 u) on this scale is defined as $^{1}/_{12}$ th of the mass of a $^{12}_{6}C$ atom. This type of atom is used because it can easily be isolated from other carbon atoms.

Summary

The **isotopes** of an element have the same number of protons and different numbers of neutrons.

$^{A}_{Z}X$ is the symbol of an isotope with Z protons and $(A–Z)$ neutrons.

Questions

1. (a) What type of charge is carried by (i) an electron, (ii) a proton?

(b) The most common isotope of oxygen has 8 electrons, 8 protons and 8 neutrons in each atom.

(i) What is the mass of this type of atom in atomic mass units?

(ii) What is the total charge of the nucleus of this type of atom?

(iii) What would be the overall charge of such an atom if it lost one electron?

2. Fill the gaps in the following paragraph using words from the word list below.

atom electron isotope molecule neutron proton

(a) Each _______ of carbon dioxide CO_2 consists of two oxygen _______ and one carbon _______ .

(b) $^{238}_{92}U$ is an _______ of uranium. Each nucleus of $^{238}_{92}U$ is composed of 92 _______ and 146 _______.

(c) The lightest atom is the hydrogen isotope $^{1}_{1}H$. This type of atom consists of one _______ and one _______ .

More about bonds

The atoms in a molecule are joined to each other by **bonds** which hold the atoms together. Atoms and molecules in liquids and solids form bonds that prevent the particles moving away from each other. If the bonds are sufficiently strong, the molecules lock each other together in a rigid structure. There are several different types of bonds. The electron arrangement of each type of atom determines the type of bond formed. The type of bond formed determines the physical state of the substance at a given temperature.

The electrons in an atom move round the nucleus like planets moving round the Sun. The electrical attraction between each electron and the nucleus prevents the electron from leaving the atom. The electrons in an atom can only occupy certain orbits round the nucleus. These allowed orbits are called *shells*. Each shell can hold up to a certain number of electrons. The innermost shell of an atom can only hold two electrons. The next shell from the nucleus can hold up to eight electrons. The nearer a shell is to the nucleus, the less the energy is of an electron in the shell. The electrons in an atom fill the shells from the innermost shell outwards. Each electron in a shell can only escape to an outer shell if it is given a certain amount of energy. The list below shows the arrangement of the electrons in the 11 lightest atoms. These occupancy rules were worked out from the chemical properties of the elements and can be explained using *quantum mechanics*.

Element	**Atomic number Z**	**Number of electrons in each shell**		
		1st shell	**2nd shell**	**3rd shell**
Hydrogen	1	1	0	0
Helium	2	2	0	0
Lithium	3	2	1	0
Beryllium	4	2	2	0
Boron	5	2	3	0
Carbon	6	2	4	0
Nitrogen	7	2	5	0
Oxygen	8	2	6	0
Fluorine	9	2	7	0
Neon	10	2	8	0
Sodium	11	2	8	1

The chemical properties of an element depend on the electron arrangement of its atoms. Because a completely empty or full shell is in a lower energy state than a partially full shell, atoms react by gaining or losing electrons to leave the atom with full shells only.

Thus an atom of sodium reacts by losing the electron in its third shell whereas an atom of fluorine reacts by accepting an extra electron into its second shell. Other types of atoms react by sharing electrons to form completely full shells.

The Periodic Table is a table of all the known elements listed in rows in order of increasing mass number. The elements in a column of the Table are referred to as a 'group' because they have similar chemical properties. This is because each type of atom in a column has the same number of electrons in a partially filled or completely filled outer shell. For example, lithium, sodium, potassium and certain heavier elements are metals that readily react because each type of atom in this group has a single electron in an outer shell. This lone electron is easily removed when the atom reacts with another atom. Another distinctive group of elements are the inert gases, helium, neon, argon, xenon and krypton. Atoms of all these electrons have full electron shells and are therefore unreactive.

Types of bonds

The type of bond formed between two atoms or molecules depends on how the atoms or molecules can lose or gain or share electrons to reach a lower energy state.

- *A covalent bond* is formed as a result of two atoms sharing a pair of electrons. Each atom contributes an electron to the bond from its outer shell. For example, hydrogen gas consists of molecules which each consist of two hydrogen atoms. The two atoms form a covalent bond by sharing their two electrons so each of the atoms in the molecule has a full shell of electrons. Another example is provided by the carbon dioxide molecule which consists of one carbon atom and two oxygen atoms. Each oxygen atom forms two covalent bonds with the carbon atom. The outer shell of the carbon atom in a molecule is therefore filled with eight electrons, four originally belonging to the carbon atom and two from each oxygen atom. The outer shell of each oxygen atom in the molecule also has eight electrons, six from the oxygen atom and two from the carbon atom.

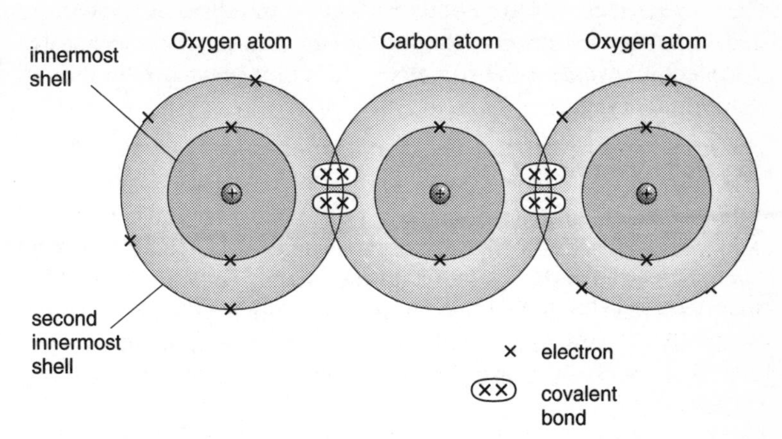

Figure 9B Covalent bonds

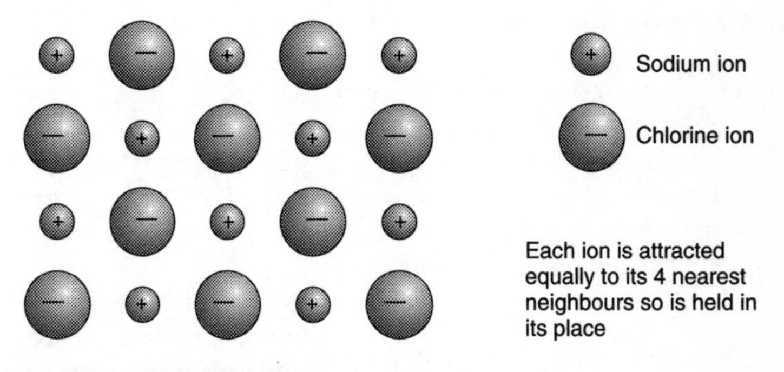

Figure 9C Ionic bonding

- An *ionic bond* is formed when an atom of one element loses an electron in its outer shell to an atom of a different element. The donor atom becomes positively charged because it loses an electron whereas the atom that gains the electron becomes negatively charged. Charged atoms are called **ions**. Hence the atoms that form ionic bonds attract each other because they are oppositely charged. The characteristic shape of a crystal is because it contains positive and negative ions arranged in a regular pattern referred to as a *lattice*. Each ion is held in place in the lattice by ionic bonds formed with adjacent oppositely charged ions. Ionic crystals dissolve in water because water molecules weaken the forces of attraction between the ions in an ionic crystal.

Crystal formation

Observe a few grains of common salt under a magnifying glass and you will see that the grains are cubic in shape. Watch the grains dissolve at the edge of a drop of water. Then blow steadily over the water to make it evaporate. You will see the grains form again. The chemical name of common salt is sodium chloride. It contains positively charged sodium ions and negatively charged chlorine ions.

- *Metallic bonds* are formed in a solid metal which consists of a lattice of positive metal ions surrounded by electrons which move about freely inside the metal. Each metal atom has lost one or more of its outer electrons. These electrons are referred to as *conduction electrons* because they carry electric charge through the metal when a voltage is placed across the metal. The conduction electrons prevent the positive ions from moving out of place in the lattice. Solid metals are much stronger than many other solids because metallic bonds are equally strong in all directions.
- *Molecular bonds* act between uncharged molecules at close range. This type of bond is formed when two molecules are so close that the electrons of each molecule are slightly attracted to the nucleus of the other molecule. Molecules in liquids move about at random but remain in the liquid because bonds between the molecules prevent them from leaving the liquid surface. If the liquid temperature is raised, the faster moving molecules near the surface can break away from the liquid and become gas molecules.

Summary

The electrons in an atom can only occupy certain allowed orbits called *shells*. Each shell can hold up to a certain number of electrons.

Types of bonds:

A covalent bond is where atoms share electrons.

An ionic bond is where one type of atom gains one or more electrons from a different type of atom.

Metallic bonds consist of a lattice of positive metal ions surrounded by electrons which move about freely inside the metal.

A molecular bond acts between two molecules that are so close that the electrons of each molecule are slightly attracted to the nucleus of the other molecule.

Questions

Q3. Explain in terms of electrons why
(a) helium atoms do not interact with other types of atoms,
(b) sodium ions carry a positive charge.

Q4. What type of bond is formed when
(a) two atoms share a pair of electrons,
(b) one type of atom gains an electron and one type loses an electron.

Solids and structure

A solid object has its own natural shape because the force bonds lock the atoms of the solid together. When a solid object is stretched or compressed or twisted, the atoms are pulled away from each other. If the forces distorting an object are removed and the object returns to its natural shape, the object is said to possess **elasticity**. If the distorting forces are sufficiently large, the object will not regain its natural shape and is permanently distorted. For example, a perspex ruler that is bent slightly becomes straight again when the bending forces are removed. If the bending forces are large enough, the ruler bends permanently or cracks. Another example is a paper clip which is designed to be strong enough to hold sheets of paper together but not too strong or it would be impossible to use. If too much force is applied to a paper clip, it bends permanently. The limit beyond which an object loses its elasticity is referred to as its *elastic limit*. Above its elastic limit, the object is permanently distorted and is said to be *plastic* rather than elastic. The bonds between atoms in the object break and are re-formed with different atoms when the object shows plastic behaviour.

Elasticity tests

Carry out each of the following tests and decide if the elastic limit has been exceeded in each case.

1. Measure the length of a rubber band using a millimetre ruler. Stretch the rubber band and then release it. Measure its unstretched length again.
2. Repeat the above test with a strip of polythene cut from a polythene bag.
3. Repeat the test with a length of string.

You should find that the rubber band and the length of string remain within their own elastic limits but the polythene does not.

Atoms in solids

A solid may be classified as either crystalline, amorphous or as a polymer.

1. Crystalline solids

The atoms in a crystal are arranged in a regular pattern. As a result, the crystal has a recognizable shape. For example, sodium chloride crystals are cubic because sodium and chlorine ions occupy alternate positions like opposing sports fans next to each other in rows of seats. Imagine each 'reds' fan sitting with an opposing 'blues' fan immediately in front, behind and on either side. The analogy ends there as the sodium ions are much smaller than the chlorine ions and carry opposite charge which is why the structure holds together. Crystals with a particular shape consist of rows of atoms arranged in such a way as to create the characteristic shape of the crystal.

A crystal is difficult to break but a small amount of force applied in a suitable direction can cause it to cleave into two smaller crystals. Such a force can be applied by tapping the crystal with a sharp edge parallel to one of its faces. The same amount of force applied differently to the crystal would have no effect on it. The reason why the crystal cleaves as above is because the tapping force is parallel to the layers of atoms in the crystal and it causes the atoms in one layer to slide over the atoms in an adjacent layer. The effect is not unlike applying a force to a stack of cards on a table so the cards slide over each other. The same force pushing down on the stack of cards would have little effect on the stack.

Metals consist of tiny crystals called **grains**, packed together with no spaces between adjacent grains. To visualize the grain structure inside a metal, imagine an aerial view of a landscape consisting of small ploughed fields with no spaces between the fields and with a different direction of furrows in each field. The atoms in a grain are arranged in rows like the furrows in a field. Across the boundary between two adjacent grains, the rows change from one direction to a different direction like the furrows in different directions in two adjacent fields. The grains in a metal can be observed using a powerful microscope if the metal surface is cleaned and polished to make it very smooth. The strength of a metal is due to the presence of grain boundaries. When forces are applied to a metal within the elastic limit of the metal, the layers of atoms in each grain are unable to slide past each other because the layers in adjacent grains are not in the same direction. Imagine many packs of cards, each held by a rubber band, jumbled together in a large cardboard box. The box would be difficult to distort because the cards that could be made to slide by a sideways push force would be prevented from doing so by adjacent packs. If all the cards in all the packs were horizontal, then the box could easily be distorted by pushing it sideways.

Testing steel

Heat treatment of steel alters its strength because it changes the size of the grains. If the end of a steel wire is heated in a flame and then suddenly cooled in water, the wire becomes very brittle and snaps easily. Heating the wire causes the grains to become larger and fewer so there are fewer boundaries in the metal. Sudden quenching in water makes these large grains permanent and therefore weaker as there are fewer grain boundaries to stop layers of atoms sliding past each other when forces are applied to the metal.

2. Amorphous solids

The atoms in an amorphous solid are locked together in contact with each other at random. In a liquid, the atoms are in contact with each other but they are not locked together and move about in random directions. An amorphous solid is sometimes described as a 'frozen liquid' as if the atoms of a liquid suddenly stopped moving and locked together haphazardly. Glass is an example of an amorphous solid, consisting of silicon atoms and oxygen atoms linked together by covalent bonds. Glass is brittle because the effects of stress in the glass due to bending forces concentrate at tiny cracks in the surface. These cracks then deepen which causes the stress in the glass to become even more concentrated so the glass snaps suddenly.

If glass is heated strongly, it loses its brittleness as cracks in the surface disappear and it softens enough for it to be blown or drawn into any desired shape by a skilled glassblower. On cooling, it sets in its new shape when it recovers its stiffness and brittleness.

3. Polymers

A polymer consists of long molecules, each consisting of a chain of atoms, formed as a result of identical shorter molecules joining together end-on. For example, polythene consists of long polyethylene molecules formed by making ethylene molecules join end-on. Each ethylene molecule consists of two carbon atoms and four hydrogen atoms. The two carbon atoms are joined together by a double bond (i.e. two covalent bonds) and two hydrogen atoms are attached to each carbon atom. When the ethylene molecules join end-to-end, the double bond is replaced by a single bond which enables the carbon atoms to link together in a long row. Each carbon atom in the polymer molecule is thus joined by a single covalent bond to a carbon atom on either side and by a covalent bond to each of two hydrogen atoms.

Many different polymers exist, all consisting of long molecules which are chains of identical shorter molecules. The long molecules of a polymer are tangled together in amorphous polymers such as polycarbonate which is used to make 'hard hats'. The atoms in each chain are linked together by covalent bonds. Bonds form between chains where two chains are in contact. These cross-links can be strong (e.g. covalent bonds) or weak (e.g. molecular bonds) depending on the type of polymer. *Thermoplastics* such as polycarbonate and amorphous polythene are polymers with weak cross-links that break when the polymer is warmed, causing it to soften. These cross-links re-form when the polymer is cooled again so the polymer can be moulded into any shape by warming then cooling it. *Thermosetting polymers* such as bakelite which is used for electrical fittings have strong cross-links that make them stiff and heat resistant.

Rubber is a natural polymer which consists of long molecules that tend to curl up. Cross-links between the molecules break when the rubber is stretched and re-form when it is released. Raw rubber is made much stiffer by adding sulphur in a process known as 'vulcanizing'. The sulphur atoms form strong cross-links between the rubber molecules thus making the rubber much stronger and stiffer.

A tyre tale

With every turn of a tyre, every part of the tread is squashed then stretched as it makes contact with the road surface. The tyre is made of rubber because rubber is resilient which means that it can be stretched and squashed repeatedly without losing its strength. Energy must be used to stretch rubber and some of this energy is recovered when the rubber 'rebounds' back to its normal shape. The unrecovered energy heats the rubber and raises its temperature. Fuel usage could be reduced if all the energy used to stretch and squash a car tyre was recoverable. However, the tyre would need to be very stiff which could make journeys very uncomfortable.

Summary

Elasticity is the property of a solid that enables it to regain its natural shape after it has been distorted.

A **crystalline** solid consists of atoms in a regular arrangement.

An **amorphous** solid consists of atoms locked together at random.

A **polymer** consists of long chain molecules.

Metals consist of tiny crystals called **grains**.

Questions

Q5. What is meant by (a) elasticity, (b) the elastic limit of an object.

Q6. State whether each of the following solid materials is crystalline, polymeric or amorphous:

a) steel, (b) polycarbonate, (c) sodium chloride, (d) glass, (e) rubber

Q7. PVC is a polymer used to make electrical wires and cables.

(a) What properties of PVC make it suitable for this use?

(b) Why could it be dangerous if a PVC cable overheats?

Molecules in fluids

Any substance that can flow is a fluid. The molecules in a fluid are not locked together and they move about at random. In a gas, the molecules move very fast and are separated by large empty spaces. The molecules in a liquid move about more slowly than in a gas and they are in contact with each other. Liquids (and solids) are much denser than gases because the atoms and molecules in liquids (and solids) are much closer to each other than they are in gases. The forces between gas molecules are too small to affect the fast-moving molecules in a gas. However, in liquids and solids, the molecules move more slowly than in a gas and the forces between them are strong enough to pull the molecules together.

Diffusion

Observe a few drops of milk spreading out in a glass of water without stirring the water. The milk gradually spreads throughout the water. This process of the spreading of a fluid in another fluid is known as **diffusion**. It occurs because the molecules of each fluid are in motion and so they move about until they are evenly distributed. Diffusion takes place with gases as well as with liquids. For example, vapour from a liquid with a strong odour spreads out through the surrounding air. Someone near the liquid would notice the odour before someone further away. Initially, the vapour molecules are concentrated near the surface of the liquid but gradually they move away from the liquid in every direction until they are evenly distributed throughout the air. This process is the same as what happens when a container is filled with red beads then with blue beads on top. If the lid is then closed and the container is shaken repeatedly, the beads gradually move about until they are distributed at random throughout the container. Provided there are a large number of beads present, the beads become evenly distributed.

Diffusion tests

Some further examples of diffusion can easily be observed in the following situations:

1. Release a drop of ink in a glass of water and observe the ink gradually spreading out. Eventually, the ink becomes evenly spread through the water.
2. Spray a small quantity of deodorant or perfume onto a cloth then walk to the other side of the room. You should be able to detect the deodorant using your nose within a few minutes.
3. Pour a small quantity of vinegar onto a saucer and you should be able to detect its presence almost immediately using your nose.

Diffusion always results in an even spread of each substance throughout the container. This happens because the molecules move about at random and there is a huge number of molecules per cubic millimetre in any liquid or gas. If there were just a few molecules per cubic millimetre on average, the distribution of molecules would be uneven and constantly changing. However, with a large number of molecules per cubic millimetre, fluctuations in the distribution would be negligible. Imagine a thousand million molecules in a container of volume 1000 cubic millimetres (= 1 cubic centimetre). Each cubic millimetre would therefore contain 1 million molecules on average at any given instant. The probability of there being significantly more or less molecules in a cubic millimetre at any instant would be negligible. Diffusion results in even spreading because the possibility of uneven spreading is negligible for large numbers of molecules.

Viscosity

Water flows more easily than cooking oil which flows more easily than syrup. The **viscosity** of a fluid is a measure of its resistance to flow. Syrup has a higher viscosity than cooking oil which has a higher viscosity than water. In general, gases flow much more easily than liquids so the viscosity of a gas is much lower than the viscosity of a liquid.

Viscosity depends on temperature. Oil flows more easily the warmer it is. The lower the temperature of the oil, the less easily it flows. This is part of the reason why starting a vehicle can be difficult on a cold morning. The oil in the engine is cold and therefore less effective as a lubricant. Some liquids become more or less viscous when stirred. For example, paint in a tin becomes 'thinner' if it is stirred enough which means that it flows more easily and is easier to apply. The opposite effect happens to wallpaper paste when it is stirred. Gradually it becomes 'thicker' and harder to stir until it reaches a consistency allowing it to be spread in a sticky paste onto the wallpaper.

Viscosity is due to the movement of molecules in a fluid between layers moving at different speeds. Molecules that transfer from slow-moving layers to faster moving layers 'drag' on the faster moving layers. Thus viscosity is a form of friction between layers in a fluid moving at different speeds. Fast-moving layers in a moving fluid are dragged by slower moving layers which are dragged by layers moving even slower and so on. For example, when a fluid flows through a pipe, the surface drags on fluid nearby which drags on fluid further away which drags on fluid further away and so on.

A health problem

Narrow arteries restrict blood flow and cause the heart to work harder to keep the blood circulating round the body. The flow rate through a pipe of diameter D is proportional to D^4. This means that if an artery is narrowed by internal deposits to half its normal width, the flow rate would be reduced to one-sixteenth of its normal rate unless the heart works 16 times harder. In reality, the heart would be unable to work that much harder so the flow rate would be reduced considerably.

Summary

Diffusion is the process of the spreading of one fluid into another fluid.

The **viscosity** of a fluid is a measure of its resistance to flow.

Questions

Q8. (a) Describe and explain what is observed when a soluble aspirin is dropped into a glass of water.

(b) A drinking glass contains some lemonade and a slice of lemon. The lemon slice is then removed. Explain why the lemonade still tastes of lemon even though the slice has been removed.

Q9. (a) Place these liquids in order of increasing viscosity:

cream syrup tar water

(b) Correction fluid is used to erase errors on paper. Why is it necessary to shake the bottle before use if it has not been used for some time?

(c) Describe a situation where a liquid becomes more viscous when it is warmed.

Pressure

A fluid will generally flow from high pressure to low pressure, driven by the force due to the pressure difference. For example, in most homes, cold water is supplied to a cold water tank via feed pipes connected to a main water pipe outdoors under the ground. Water is pumped into the main pipe at high pressure at a local pumping station and flows along the pipe into the homes when required.

Pressure is caused when an object exerts a force on a fluid. The fluid in a syringe squirts out of the nozzle of the syringe when a force is applied to the end of the syringe. The greater the force, the faster the fluid squirts out. The narrower the syringe, the easier it is to use because less force is needed to create a certain pressure with a narrow syringe compared to a wide syringe. The pressure in the syringe is the force per unit area acting on the fluid. The greater the force or the smaller the area of the object, the higher the pressure created in the fluid.

Pressure = force per unit area

The unit of pressure is the *pascal* (Pa), which is equal to the pressure exerted by a force of 1 newton acting on an area of 1 square metre perpendicular to the area.

Hydrostatic pressure

The force of the Earth's gravity causes pressure in a fluid. The pressure in a liquid in an open container increases with depth below the surface. This is why water runs out of a sink more slowly as the sink empties. The pressure of the water at the plughole depends on the depth of water in the sink. As the depth becomes less, the pressure becomes less and the rate of flow drops.

Consider a liquid of density ρ in an upright cylinder open at the top, as in Fig. 9D.

1. The weight of the liquid in the cylinder = mg, where m is the mass of the liquid in the cylinder.

2. The pressure on the base of the cylinder

$$= \frac{\text{the weight of the liquid}}{\text{the base area } A} = \frac{mg}{A}.$$

3. The volume V of the liquid in the cylinder = Ah, where h is the height of the liquid in the cylinder.

The pressure at the base $= \dfrac{mgh}{Ah} = \dfrac{mgh}{V} = \rho gh$

since the density of the liquid $\rho = \dfrac{m}{V}$.

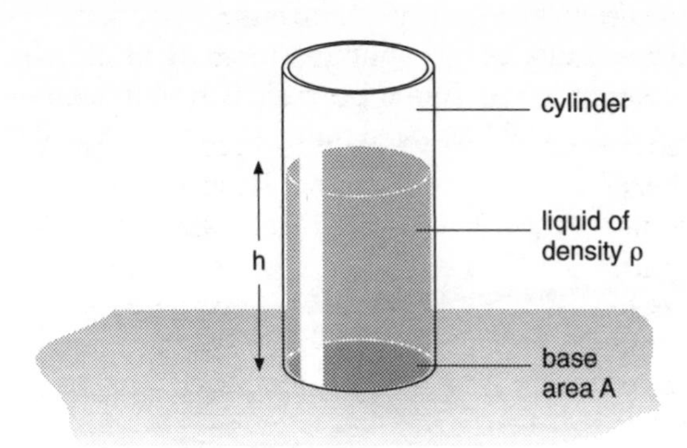

Figure 9D Hydrostatic pressure

Thus the pressure, p, due to a column of liquid of height h and density ρ is given by the equation,

$$p = \rho g h$$

Worked example

$g = 9.8$ m/s^{-2}

Calculate the pressure due to a depth of 5.0 metres in water. The density of water = 1000 kg m^{-3}

Solution

$p = h\rho g = 5.0 \times 1000 \times 9.8 = 49\ 000$ Pa.

About high blood pressure

Blood pressure is traditionally measured using an instrument called a *sphygmomanometer* which consists of an inflatable cuff attached to a hand pump and a mercury-filled container. The cuff is wrapped round the upper arm and inflated so it cuts off the blood supply to the lower arm. The pressure in the cuff causes the mercury to rise from the container up a vertical glass tube fixed to the container. The cuff pressure is then released and the mercury level drops. The pulse at the wrist can be detected when the cuff pressure drops sufficiently to allow blood flow to resume. The height of the mercury in the glass tube at this point is a measure of the maximum or 'systolic' blood pressure. Blood pressure is usually expressed in millimetres of mercury (abbreviated mmHg as the chemical symbol for mercury is Hg).

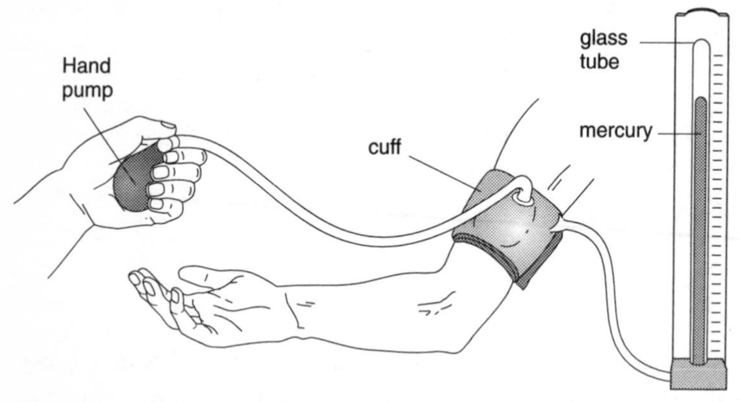

Figure 9E Measuring blood pressure

Blood pressure varies due to the beating of the heart. A person's blood pressure is usually measured and recorded as the systolic pressure and a lower value known as the 'diastolic pressure'. The blood pressure of a typical young person is 120 / 80 mmHg which means the systolic pressure is 120 mmHg and the diastolic pressure is 80 mm Hg. A systolic pressure reading considerably in excess of this indicates high blood pressure and the need for medical treatment.

To convert mmHg to pascals, use the formula $p = h \rho g$ where h is the reading converted from mm to metres, ρ is the density of mercury which is 13600 kg m^{-3} and g = 9.8m/s^{-2}.

For a reading of 120 mmHg, $h = 0.120$ metres,

$\therefore p = 0.120 \times 13600 \times 9.8 = 16000$ Pa.

Gas pressure

The pressure of a gas is due to gas molecules hitting the container surface and rebounding. In a gas, the average separation of the molecules is much larger than the size of a molecule. In comparison, the molecules in a liquid or in a solid are always in contact with each other. Gas molecules move about at high speeds, colliding with each other and the container at random. The motion of gas molecules in a box is like squash balls flying about in a squash court, hitting the walls and rebounding and occasionally hitting each other. The pressure due to the effect of gravity on the gas is much much smaller than the pressure due to the repeated impacts of the molecules on the container. The force due to the impacts is smoothed out because the number of molecules per second hitting the surface of the container is very large.

The first direct evidence for the existence of molecules was obtained by the eighteenth-century Scottish botanist, Robert Brown. He used a microscope to observe tiny pollen grains floating in air and he saw that the grains quivered as they moved about at random. He reasoned that their quivering haphazard motion was because each grain was bombarded unevenly by fast-moving molecules. The uneven impacts by molecules too small to see were forceful enough to push each tiny grain about at random.

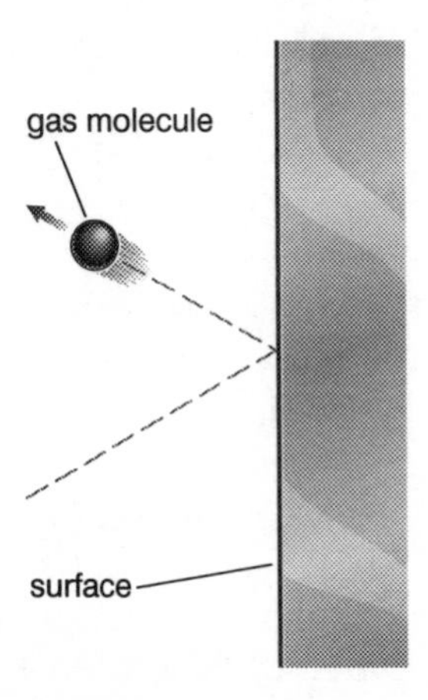

Figure 9F Gas pressure

The pressure of a gas in a sealed container increases if the gas is heated. This happens because the molecules of the gas move about faster when the gas is heated. Thus the impacts on the container surface are more forceful and more frequent so the pressure of the gas is greater. If the gas is cooled, the pressure falls as the molecules move slower and their impacts are less forceful and less frequent. The pressure would be zero if the gas was cooled to absolute zero which is –273°C. See p. 63.

The pressure of a gas at constant temperature increases if the volume of the gas container is reduced. For example, if the air in a cycle pump is trapped by blocking the outlet then compressed by pushing the handle down the barrel, the pressure of the air in the pump increases. The reason why this happens is that the air molecules hit the container surface more often in a smaller space because they have less distance to travel between successive impacts with the container surface.

Summary

Pressure = force per unit area acting perpendicular to the area. The unit of pressure is the pascal (Pa), which is equal to 1 newton per square metre.

Pressure due to a column of liquid = $\rho g h$ where h is the height of the column and ρ is density of the liquid

Gas pressure: The pressure of a gas is due to gas molecules hitting the container surface and rebounding.

Questions

Q10. In terms of pressure, explain why:

(a) air supplied to a diver under water must be at high pressure,

(b) expanding your chest causes you to draw air into your lungs,

(c) the wall of a dam is thicker at the base than at the water surface.

Q11. (a) Explain why the pressure in a vehicle tyre is greater immediately after a car journey than before.

(b) The brake system of a vehicle contains a brake fluid. Explain why it is important that the brake system contains no air bubbles.

Q12. (a) Calculate the hydrostatic pressure at the bottom of a swimming pool of depth 2.0 m. The density of water is 1000 kg / m^3.

(b) A blood pressure monitor attached to a patient gives a reading of 120 mmHg. Convert this reading to pascals. The density of mercury is 13 600 kg / m^3.

10 THE AGE OF NEW PHYSICS

This chapter is about the two great theories of modern physics, the quantum theory and the theory of relativity. Both theories, undiscovered at the end of the nineteenth century, revolutionized physics in the twentieth century. The quantum theory is about observations, processes and interactions involving events at a sub-microscopic scale. Information technology and communications would not have developed without the quantum theory which provides the theoretical basis of devices such as transistors and integrated circuits. Relativity theory is about the nature of space, time, mass and energy. Nuclear power and discoveries such as quarks and black holes are consequences of the theory of relativity. In this chapter, we will look at the ideas of quantum theory and the theory of relativity before meeting nuclear power, quarks, black holes and other big discoveries in the final chapters of this book.

Quantum theory

The new discoveries in physics at the start of the twentieth century created a revolution in the subject as previous assumptions about the nature of matter and radiation were overthrown. The discovery of photoelectricity (see p. 130) led to the conclusion that light consists of wavepackets or 'photons'. The smallest quantity of light is therefore the photon. Thus a light beam consists of many photons. The discovery of the electron produced the conclusion that the smallest quantity of electric charge is the charge carried by the electron. An electric current consists of many electrons moving on average in the same direction. Quantities such as electric charge and light were previously thought of as quantities that flow or spread out continuously. However, the new physics showed that these quantities are multiples of basic amounts, the basic amount referred to as a *quantum*. Thus the quantum of charge is the charge carried by the electron. The quantum of light is the photon. These ideas collectively became known as the *quantum theory*. In the next few pages, we will look at some of the key rules of quantum theory.

Wave mechanics

We have seen that light has a wave-like nature because it produces an interference pattern when it passes through two closely spaced slits. Light also has a particle-like nature because it consists of wavepackets or photons. If light can have a wave-like nature and a particle-like nature, can a matter particle such as an electron possess a wave-like nature? This question was considered by the French physicist, Louis de Broglie, in 1923 who put forward the hypothesis that particles have a dual nature which is wave-like or particle-like according to the situation which the particle is in. He proposed that the wavelength, λ, of the waves and the speed, υ, of the particle are linked through the equation

$$\lambda = \frac{h}{m\upsilon}$$

where h is the Planck constant and m is the mass of the particle.

Within a few years, de Broglie's strange ideas had been shown to be true when it was discovered that a beam of electrons directed at a thin metal foil is scattered into certain directions only. The same type of effect happens when a beam of light is directed at a set of closely spaced slits; light spreads out on passing through each slit. This effect, known as **diffraction**, happens when any type of wave passes through a gap. The diffracted light waves reinforce in certain directions only. Diffraction is a wave property and reinforcement is an example of interference (see p. 122) which is also a wave property. Electrons in a beam thus behave as waves when the beam passes through a thin metal foil because they emerge in certain directions only, just as a light beam does when aimed at a set of parallel slits.

The *electron microscope* is much more powerful than a light microscope. Electrons in a beam in a vacuum tube pass through a thin specimen and are scattered by it. Magnetic 'lenses' are used to focus the scattered electrons onto a fluorescent screen where they

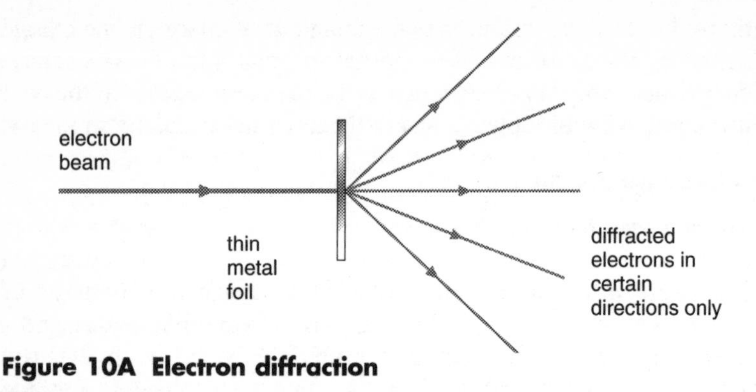

Figure 10A Electron diffraction

form a magnified image of the specimen. Much more detail can be seen in the image than in a light microscope. The amount of detail possible in a microscope depends on how much diffraction occurs when the waves pass through the lenses. The smaller the wavelength of the waves or the bigger the gap they must pass through, the less the diffraction that occurs and the greater the detail that can be seen in the image. The electrons in the beam have a much shorter wavelength than light waves so much more detail is seen.

Energy levels and line spectra

Wave mechanics is used to explain why the electrons in an atom occupy 'shells' and why an electron in an atom has a fixed amount of energy which depends on the shell it occupies. The energy of an electron in a shell is called an *energy level*. An electron trapped in an atom does not have enough kinetic energy to escape from the atom. The force of electrostatic attraction between the electron and the nucleus prevents the electron from leaving the atom. In effect, the electron is trapped in a 'well'. The de Broglie wavelength of an electron in the well must be such that a whole number of wavelengths must fit across the well. Because the speed of an electron and its de Broglie wavelength are related by the Broglie equation $\lambda = h / mv$, it follows that the speed of an electron in a well can only have certain values. Thus the kinetic energy of the electron can only have certain values. As a result, the energy of the electron in the well is at certain levels only. The exact values for these levels depends on the shape and depth of the well which depends on the charge of the nucleus and how many other electrons are present. The key point however is that the energy of an electron in a confined space is 'quantized' not continuous. Each energy level corresponds to a 'shell' surrounding the nucleus where an electron at that energy level is located.

The hydrogen atom is the simplest atom as it consists of a single proton as the nucleus and a single electron. The ideas above give the energy levels of the hydrogen atom at values of E_1 / n^2, where $n = 1$ for the innermost shell, $n = 2$ for the next shell, etc. and E_1 is the energy level of the innermost shell.

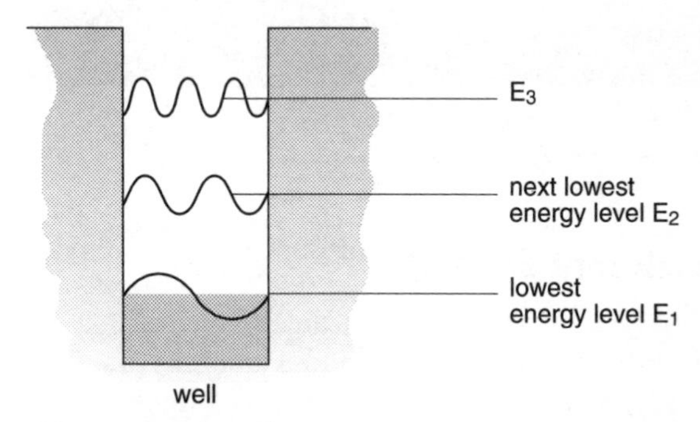

Figure 10B Energy levels

The energy levels of an atom determine the *line spectrum* of light emitted by an atom. A line spectrum consists of a pattern of well-defined bright coloured lines, each line being due to light of a certain wavelength from the atom. When an atom emits light, an electron in a shell moves to a different shell at a lower energy level. The electron loses energy and releases a photon which carries away the energy lost by the electron. Photons with the same energy all have the same wavelength so the spectrum of light from this type of atom consists of well-defined wavelengths. The spectrum of light can be seen by using a prism to split the light into coloured lines, each line being due to light of a certain wavelength. The pattern of coloured lines is the line spectrum of the type of atom that emits the light. Each type of atom produces its own line spectrum because the energy levels of each type of atom are unique to that type of atom. The hydrogen spectrum consists of lines that match exactly electron transitions between energy levels given by the formula above.

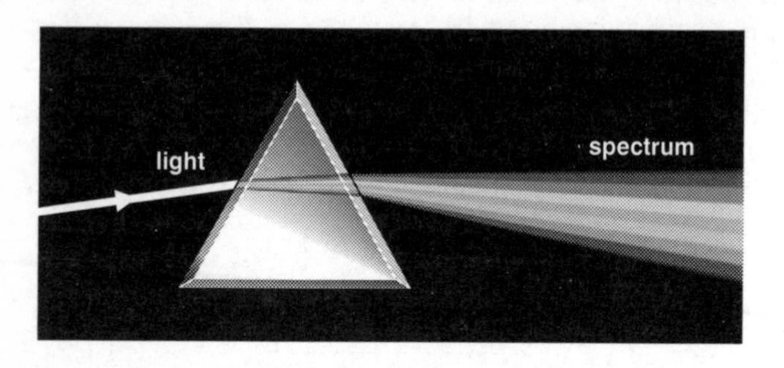

Figure 10C A line spectrum

The Uncertainty Principle

So far we have seen that quantum theory means that energy, the capacity to do work, is not a continuous quantity but is quantized. Everyday experience suggests otherwise as a light ray seems to be a steady stream of energy and the speed of any everyday moving object is not restricted to allowed values only. Yet particles at the sub-atomic scale must obey the rules of the quantum world. The reason why we do not experience the quantum world in our everyday experiences is because such experiences are due to very large numbers of particles, each with its own energy. We fail to see individual photons of light because even the faintest image seen by the eye is due to huge numbers of photons.

Quantum effects are observed where the action of making an observation affects the outcome of the observation. For example, when a thermometer is used to measure the temperature of an object, energy transfer from the object to the thermometer takes place. For a large object, this transfer makes no measurable difference to the temperature of the object. However, if the object is sufficiently small, the action of making the measurement would alter the object's temperature. Another example is the determination of the speed of an object. This could be done by timing the object as it passes two fixed positions. The object would need to be observed as it passes each position. On a large scale, the act of observing the object would not affect its speed. However, if the object is a sub-atomic particle, photons used to observe it could change its speed when they are scattered by it. Thus the action of making an observation on this scale determines what is observed.

Quantum effects are thus uncertain and need to be described as such. These ideas were put on a formal basis by the German physicist Werner Heisenberg who proved that there is a fundamental uncertainty in the state of a particle in that its speed and its position cannot be determined simultaneously. The less uncertain the position of a particle is, the more uncertain its speed is. The less uncertain the speed of a particle is, the more uncertain its position is. Heisenberg's **Uncertainty Principle** leads to the conclusion that the shorter the duration of a process, the more uncertain the energy of the particles involved. Clearly, energy in the quantum world is a much more mysterious quantity than in the world of machines and engines.

Summary

The **de Broglie wavelength** of a particle, $\lambda = \frac{h}{m\upsilon}$

where h is the Planck constant, m is the mass of the particle and υ is its speed.

The Uncertainty Principle: The less uncertain the position of a particle is, the more uncertain its speed is. The less uncertain the speed of a particle is, the more uncertain its position is.

Questions

Q1. In each of the following experiments, state whether the electron behaves as a wave or a particle.

(a) An electron in a beam is deflected by a magnetic field.

(b) An electron in a beam passes through a metal foil and is diffracted in a certain direction.

(c) An electron in an atom is knocked out of the atom by a photon.

Q2. (a) Describe the difference between a continuous spectrum and a line spectrum.

(b) Explain why an optical line spectrum indicates that the electrons in the light source are at well-defined energy levels.

Q3. The diagram below shows three of the energy levels of the electrons in a certain atom.

(a) List the three different downward electron transitions that are possible in this atom.

(b) Which of the three electron transitions produces the smallest photon energy?

Q4. The speed of the electrons in an electron microscope can be increased by increasing the voltage applied to the microscope. The voltage is increased so the speed of the electrons is doubled.

(a) How is the de Broglie wavelength of the electrons in the beam changed?

(b) How is the final image affected by this increase of speed?

Relativity

No object can be made to travel faster than light. The speed of light in a vacuum is the ultimate speed limit. Albert Einstein was the first person to realize the fundamental significance of the speed of light. Before Einstein, scientists had attempted to find out if the speed of light was affected by the speed of the observer or the light source. Newton's laws of motion predicted that if an object is thrown forwards at a speed of 10 m/s from a vehicle moving in a straight line at 30 m/s, the speed of the object relative to the ground is 40 m/s. Physicists before Einstein thought the same rule should apply to light from a moving light source. They knew that the speed of light in a vacuum is 300 000 km/s so they reasoned that a light beam on a moving source aimed in the forward direction of the source ought to travel faster than 300 000 km/s. Experiments were devised to test this prediction but the speed of light was found to be the same, regardless of the speed of the light source. No one could explain this result satisfactorily until Einstein provided the answer in 1905 when he published the theory of relativity.

Einstein at that time was no more than a junior scientist at the Swiss Patent Office. His official duties were not too demanding and he was able to develop his ideas in his spare time. His theory of relativity started with just two assumptions:

1. The speed of light is the same for any observer, regardless of the speed of the source or the speed of the observer.
2. All motion is relative.

The first assumption cuts out the argument about whether the speed of light depends on the speed of the light source. The second assumption means that all physical laws expressed in mathematical form must be the same for all observers moving at constant velocity relative to each other. Einstein put these assumptions forward as his starting points and then worked mathematically through their consequences. We shall leave aside the mathematical details here and look at why the predictions made by Einstein revolutionized our understanding of space and time.

- The observed length of a moving rod is shorter the faster the rod travels towards or away from the observer.
- A moving clock runs slower than a stationary clock.

Both effects require objects to be moving at speeds approaching the speed of light. Experiments using particle beams at speeds approaching the speed of light have confirmed Einstein's predictions. For example, the lifetime of fast-moving unstable particles called muons is longer than if they are stationary. Another example is a twin who travels at high speed away from the Earth and back again. On return, he or she would find that he or she is younger than the stay-at-home twin because a traveller's clock runs slower than a stationary clock.

Two further predictions made by Einstein in his 1905 Theory of Relativity revolutionized our understanding of mass and energy.

- The mass of a moving object increases with its speed. If an object is accelerated to a greater speed, its mass increases. Einstein showed that the mass would be infinite at the speed of light hence the conclusion that no object can be made to travel as fast as light.
- Energy E can be converted into mass m and vice versa in accordance with Einstein's famous equation

  $$E = m c^2$$

  where c is the speed of light in a vacuum. For example, the conversion of 4 tonnes (= 4000 kg) of matter into energy would release about 4×10^{20} J of energy, enough to meet the annual demand for energy for the entire world. Check the calculation yourself, given $m = 4000$ kg and $c = 3 \times 10^8$ m/s.

Albert Einstein 1879–1955

Einstein did not distinguish himself academically at secondary school in Munich. His family moved to Munich from Ulm a year after his birth and moved to Milan 14 years later. Einstein disliked the regimented school system that prevailed in Germany at that time and he left school prematurely. His family recognized his unusual talents in mathematics and physics and arranged for him to complete his secondary education in Zurich. In 1896, he gained a place to study at the Swiss Federal Institute of Technology (known as the E.T.H.) in Zurich and graduated with a physics degree four years later. He developed a reputation for asking awkward questions which his professors could not always answer. Two years after leaving E.T.H., he managed to secure a post in the Swiss Patent Office at Berne. He worked at his ideas on physics in his spare time with the support of his wife and his friends. In 1905, he published papers on the photon theory of radiation and the theory of relativity, now referred to as special relativity. He was appointed to a part-time post at the University of Berne in 1907 and became a professor in 1909. In 1914, after a brief spell in Prague, Einstein took up the post of Director of the Kaiser Wilhelm Institute for Physics in Berlin. He published the General Theory of Relativity in 1916 in which he showed that gravity is a feature of space and time. He became a household name when his prediction that gravity bends space was confirmed in 1919 by the British astronomer, Sir Arthur Eddington. In 1933, Einstein was forced to emigrate from Germany and he spent the rest of his life in America.

Experimental evidence in support of the theory of special relativity was found in the discovery that the specific charge of the electron (i.e. its charge / mass value) was smaller the greater the speed of the electrons. Further tests showed that the mass of an electron did indeed increase with speed as Einstein predicted. The masses of atoms were measured accurately and it was discovered that mass was lost when a radioactive change took place and that the energy released matched the mass loss in accordance with $E = m c^2$. More dramatically, nuclei can be made to release energy and this process can be caused on a scale that could wipe out the human race if ever nuclear weapons were to be used in a war. In addition, physicists have discovered that high energy photons are capable of turning into matter in the form of particles and antiparticles. The reverse process is also possible where a particle and an antiparticle annihilate each other and turn into photons. All these processes require the use of $E = m c^2$. Perhaps the most amazing aspect of Einstein's work is that after a century of using $E = m c^2$, the mechanism behind the conversion of mass and energy is still not clearly understood. With these thoughts in mind, we will move on in the next chapter to look at what is now known about matter and its properties at an even smaller scale than the nucleus.

Summary

Einstein's theory of special relativity assumes that

1. The speed of light is the same for any observer, regardless of the speed of the source or the speed of the observer.
2. All motion is relative.

Energy and mass: energy E can be converted into mass m and vice versa on a scale given by $E = m c^2$, where c is the speed of light in a vacuum.

Questions

Q5. (a) An atom of mass 2×10^{-28} kg releases a photon of light of energy 1×10^{-19}J. Show that the mass loss of the atom is negligible.

(b) A nucleus in an atom of mass 2×10^{-28} kg releases a photon of energy 1×10^{-13} J. Show that the mass of the nucleus decreases by about 0.5% as a result of this process.

Q6. The Sun radiates energy at a rate of 4×10^{26} watts. Calculate the amount of mass converted into energy by the Sun every second.

11 THE STRUCTURE OF MATTER

The search to discover the fundamental structure of matter has been the central theme of science for the past few centuries and is a quest that continues today. New discoveries build on previous discoveries and existing theories, enabling existing theories to be refined or new ideas developed. The ideas about atoms that were first put forward thousands of years ago by Democritus in Ancient Greece were rediscovered by John Dalton in the early nineteenth century. Dalton's theory that atoms are indivisible and indestructible worked well for almost a century until the discovery that atoms contained much lighter negative particles which became known as electrons. Experiments by Ernest Rutherford in the first decades of the twentieth century showed that every atom contains a positive nucleus where most of its mass is located. Further experiments led to the conclusion that the nucleus is composed of protons and neutrons, and that the electrons move round the nucleus. In this chapter, we will look at the discoveries that led to this picture of the atom before moving on to look at the astonishing picture revealed by physicists probing matter on incredibly small scales. We know now that protons and neutrons are composed of particles called quarks. We know that for every type of particle, there is a corresponding type of antiparticle. Like Pandora's box which has a box inside a box inside a box and so on, the search to discover the structure of matter has uncovered structure at successively smaller scales. Now the search is on to find out if quarks and electrons are manifestations of nature on an even smaller scale.

Inside the atom

New discoveries often lead to new inventions which then lead to more new discoveries. For example, in the mid-nineteenth century, the discovery of electromagnetic induction by Michael Faraday led to the invention of a range of electromagnetic devices such as the dynamo (see p. 104), the transformer (see p. 106) and the induction coil. The induction coil is like a transformer except the primary coil is connected to a battery in series with a 'make and break' switch. In operation, the 'make and break' switch repeatedly switches the current through the primary coil on and off. Each time the primary current is switched on or off, the changing magnetism through the secondary coil causes an induced voltage across the secondary coil. The secondary coil has many more windings on it than the primary coil has. Consequently, a very large voltage is induced in the secondary coil each time the make and break switch opens or closes.

The induction coil was used by Sir William Crookes in the mid-nineteenth century to investigate if gases conduct electricity at very low pressures and high voltage. Crookes found that gases conduct when subjected to high voltage at very low pressures. More importantly, he found that light is emitted by the gas in these conditions. Neon tubes and other colourful illuminated displays are the result of Crookes' discovery that the colour of light emitted depends on the type of gas in the tube. Crookes carried out further research into the nature of the particles responsible for carrying electricity in the gas. It was already known that conduction in a liquid solution such as salt water is due to the presence of positive and negative ions in the solution. When two electrodes connected to a battery are placed in the solution, the positive ions are attracted to the negative electrode and the negative ions are attracted to the positive electrode. Crookes worked out that a gas at low pressure conducts for the same reason, namely the presence of positive and negative ions in the gas. However, unlike in a liquid solution, ions in a gas are only created when the gas is at low pressure and a high voltage exists across the gas. The high voltage across the gas causes electrons to be torn out of individual gas atoms, leaving the atoms as positive ions to be attracted towards the negative electrode in the gas and the electrons to be attracted towards the positive electrode. Crookes realized that positive and negative ions were created in the gas and he reasoned that the emission of light by the gas was when a positive ion and a negative ion collide and recombine to form an atom again.

Reminder about ions

An ion is a charged atom. A positive ion is formed by removing an electron from an uncharged atom. A negative ion is formed by adding an electron to an uncharged atom.

The discovery of the electron

Research into the conduction of gases at very low pressure by Joseph Thomson led to the discovery of the electron in 1897. Thomson found that the negative ions were always the same, regardless of which gas was present. He discovered that the positive ions were gas atoms with electrons removed. Thomson made these discoveries with apparatus he designed and constructed that produced beams of ions. Thomson's apparatus was in effect the very first TV-like tube as the tube widened out into a screen with a flat end. The screen was coated internally with a chemical that produced a spot of light where the ion beam hit the screen.

By applying an electric field or a magnetic field to the beam before it hit the screen, the ions could be deflected and the deflection could be measured from the shift of position of the spot of light on the screen.

1. The electric field was created between two metal 'deflecting' plates one above the other. One plate was connected via a switch to the positive terminal of a high voltage supply unit and the other plate to the negative terminal of the supply. The deflecting plates were positioned so that the beam passed between them. When the switch was closed, the spot on the screen was seen to deflect because the ions were attracted to the oppositely charged deflecting plate.
2. The magnetic field was created by placing two coils of insulated wire either side of the narrow part of the tube and passing a current through the two coils. The deflection could be reversed by reversing the current or it could be increased by increasing the current. This principle is used in all modern TV tubes and VDUs to make an electron beam scan the screen and form a picture on the screen.

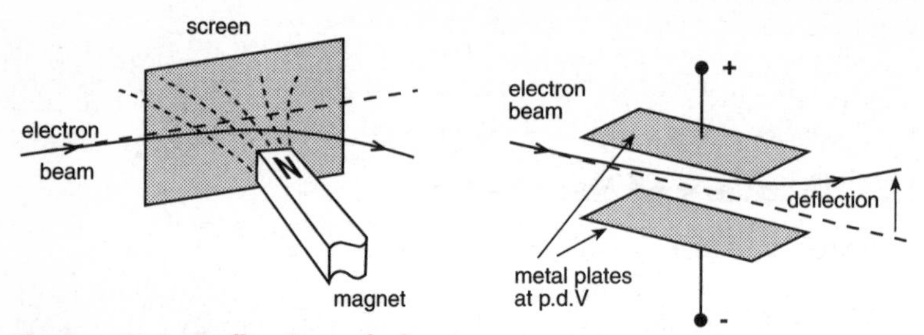

Figure 11A Deflection of electrons

Thomson found that the deflection of the negative ions did not depend on the type of gas used whereas the deflection of the positive ions did. He realized that these negative particles are in every type of atom and referred to them as 'corpuscles of electricity' before they became known as electrons. He measured the deflection of the electrons in a beam by deflecting them with electric and magnetic fields of known strengths. From his measurements, he worked out the charge / mass value of the electron (known technically as its 'specific charge'). This was a significant measurement because the specific charge of most ions had been measured decades earlier in electrolysis experiments. For example, it was known that hydrogen gas was released when electricity is passed through acidified water and that 96000 coulombs of electric charge was needed to produce 1 gram of hydrogen gas. The specific charge of the hydrogen ion was therefore known to be 96000 coulombs per gram, larger than the specific charge of any other ion until Thomson discovered that the electron has a specific charge about 1850 times greater than that of the hydrogen ion. Thomson could not be certain if this very large value is because the electron is much lighter than the hydrogen ion or because the electron carries much more charge than a hydrogen ion. However, assuming that the hydrogen ion and the electron carry equal and opposite amounts of charge, Thomson and most of his contemporaries reckoned that the electron must be much lighter than the hydrogen atom.

The charge of the electron, e, was measured by the American physicist, Robert Millikan, in 1915. His value of e was very close to the now-accepted value of 1.6×10^{-19} coulombs. The mass of the electron could then be calculated from its specific charge $e/m = 1.8 \times 10^{8}$ coulombs per gram (= 1850×96000 coulombs per gram) and its actual charge $e = 1.6 \times 10^{-19}$ C. Prove for yourself that these figures give a value of 9×10^{-28} grams (= 9×10^{-31} kilograms) for the mass of the electron.

Within 20 years of Thomson's discovery, scientists had worked out that the electrons in an atom orbit a nucleus of positive charge in the atom. The discovery of the nucleus is part of the next thread in the development of our understanding of the structure of the atom.

Summary

Gases conduct when subjected to high voltage at very low pressures.

A positive ion is formed by removing an electron from an uncharged atom. A negative ion is formed by adding an electron to an uncharged atom.

The specific charge of a particle is its charge / mass value.

Questions

Q1. Complete the following sentences using words from the list below:

atoms *ions* *light* *pressure* *voltage*

A gas conducts electricity if it is at low _______ and a high _______ is applied across it. Pairs of positive _______ and a negative _______ are created from gas _______ . The _______ move to the oppositely charged electrode in the tube. Light is emitted if a positive _______ and a negative _______ collide and recombine.

Q2. A beam of electrons hits a screen and produces a spot of light. The beam is then deflected using a magnetic field which causes the spot to move a centimetre downwards on the screen. How would the spot be affected if the magnetic field was (a) increased in strength, (b) reversed in direction?

Q3. The specific charge of the electron is 1.76×10^{11} coulombs per kilogram. The charge of the electron is 1.60×10^{-19} C. Use this data to show that the mass of the electron is 9.1×10^{-31} kg.

Radioactivity

Radioactivity was discovered accidently in Paris by Henri Becquerel in 1896 when he was conducting research into the effects of X-rays on uranium compounds. He had discovered that certain substances exposed to X-rays glow and continued to glow when the X-ray machine was switched off. He wanted to know if the reverse effect was possible, namely emission of X-rays after the substance had been exposed to strong sunlight. In readiness for a sunny day, he placed a wrapped photographic plate in a drawer with a small quantity of the uranium compound on it. After several dull days, he decided to develop the plates, expecting to observe no more than a faint image of the compound. He was therefore very surprised when he found a very strong image on the plate. He realized that the uranium compound was emitting some form of radiation without having been exposed to sunlight. Further tests showed that the substance emits this radiation continuously even when stored in darkness for long periods and that the radiation passes through glass but not through metal. The substance was described as 'radioactive' because it did not need to be supplied with energy to make it emit radiation and was therefore emitting radiation actively. Becquerel continued his research on X-rays and passed the investigation of radioactivity on to his research student, Marie Curie. Within two years, Marie Curie and her husband Pierre had discovered other substances which are radioactive, including two new elements, radium and polonium. Becquerel and the Curies were awarded the 1903 Nobel Prize in physics for their discoveries.

The nature of radioactivity was established by Ernest Rutherford who showed that the radiation is produced when unstable atoms disintegrate. He used the radiation to probe the atom and he deduced that every atom contains a positively charged nucleus where most of its mass is located. Further research showed that the nucleus itself is composed of neutrons and protons. In recent decades, scientists have discovered that protons and neutrons are composed of smaller particles which are referred to as quarks. Many questions about the sub-atomic world remain unanswered and more questions arise as more discoveries are made. We will look at some of these questions later in this chapter after we have studied radioactivity in more detail.

Radioactivity is due to the instability of a nucleus which has too many protons or neutrons. Such a nucleus becomes stable or less unstable by emitting one of three types of radiation:

1 *Alpha radiation* (α) consists of particles, each composed of two protons and two neutrons. An α particle is emitted by a very large unstable nucleus. Alpha radiation
   - is easily stopped by cardboard or thin metal,
   - has a range in air of no more than a few centimetres,
   - ionizes air molecules much more strongly than the other two types of radioactive radiation.

2 *Beta radiation* (β) consists of electrons, each emitted when a nucleus with too many neutrons disintegrates. A neutron in such a nucleus suddenly and unexpectedly changes to a proton; in the process, an electron is created and instantly emitted from the nucleus. Beta radiation
   - is stopped by 5–10 mm of metal,
   - has a range in air of about 1 metre,
   - ionizes air molecules less strongly than α radiation.

3 *Gamma radiation* (γ) consists of high energy photons. A photon is a packet of electromagnetic waves. A gamma photon is emitted from a nucleus with surplus energy after it has emitted an α or a β particle. Gamma radiation
   - is stopped only by several centimetres of lead,
   - has an infinite range in air,
   - ionizes air molecules very weakly.

The *geiger counter* is used to detect radioactive radiation. This device was invented by Hans Geiger who worked with Rutherford. It consists of a tube connected to an electronic counter. Each time an ionizing particle enters the tube, the counter registers it as a single count and a click is heard. The tube itself is sealed and hollow, with a thin window over one end and a metal rod along its

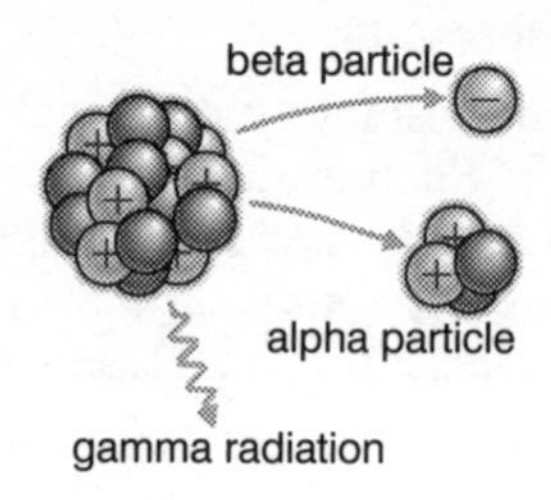

Figure 11B Radioactivity

axis. The tube contains gas at very low pressure. With several hundred volts between the rod and the tube, an ionizing particle entering the tube ionizes the gas atoms which then ionize more atoms, etc. The gas becomes conducting for a fraction of a second, causing a tiny pulse of electricity to pass through the counter and be registered.

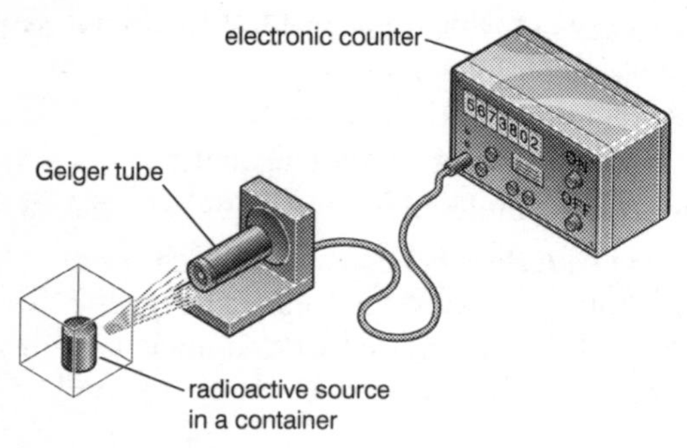

Figure 11C Using a geiger counter

To use a geiger counter:

1. the counter is set to zero.
2. the 'start count' switch is pressed and a stopwatch is started at the same time,
3. the 'stop count' switch is pressed after a measured time (e.g. 100 seconds).

The count rate is the number of counts per second. This is calculated by dividing the number of counts by the counting time. In practice, the measurement is repeated several times to obtain an average value for the count rate.

Background radioactivity must be taken into account when measuring the count rate due to a radioactive source. Background radioactivity is caused by radioactive substances present in rocks, by the effect of cosmic radiation on the atmosphere and by radioactive pollutants released into the atmosphere such as happened at Chernobyl in 1986. To take account of background radioactivity, the count rate is measured with the source present and without the source present. The count rate due to the source is the difference between the two measurements.

Radioactivity at work

1. Radioactive tracers are used in medicine and environmental technology. The tracer needs to be a radioactive isotope that emits β or γ radiation because this type of radiation can pass through materials. For example, suppose a gas leak occurs in an underground gas pipe. The leak can be pinpointed by injecting a small quantity of a radioactive gas into the pipe and then moving a geiger counter at ground level along the pipe. The source of the leak would be where the counter reads more than the background count rate. Another example from medical physics is in the diagnosis of a blocked kidney. The patient is given a drink of water containing a very small quantity of a radioactive tracer. A geiger counter is held against each kidney. For a normal kidney, the count rate would rise then fall as the water passed through. In a blocked kidney, the count rate would rise and not fall back as the water would not pass through the kidney.
2. Radioactivity is used to monitor industrial processes such as thickness monitoring of metal plating produced in a rolling mill. Molten metal is forced between two steel rollers and the metal cools to form a metal plate as it passes through. With a beta source and a geiger tube either side of the plate, the count rate varies according to the thickness of the plate. An increase of thickness would cause the count rate to drop and a decrease of thickness would cause the count rate to rise. The count rate signal is used to control the pressure on the steel rollers.

Half life

The **half life** of a radioactive isotope is the time taken for half the number of atoms of the isotope to disintegrate. Suppose 10000 atoms of a certain radioactive isotope X are present initially. The number of atoms decreases

- from 10 000 to 5000 after one half life, then
- from 5000 to 2500 after a further half life, then
- from 2500 to 1250 after a further half life, etc.

The amount of the radioactive isotope therefore decreases with time as shown in Fig. 11D which is a half-life curve. Half-life values range from a fraction of a second to billions of years. For example, the half life of uranium 238 ($^{238}_{92}U$) is about 4.5 billion years.

Radioactive disintegration is a random process. For a large number of atoms of a given radioactive isotope, the proportion that disintegrate per second is constant. This follows because of the random nature of radioactive disintegration. To appreciate this, suppose a thousand dice are rolled and all those that show a '1' are

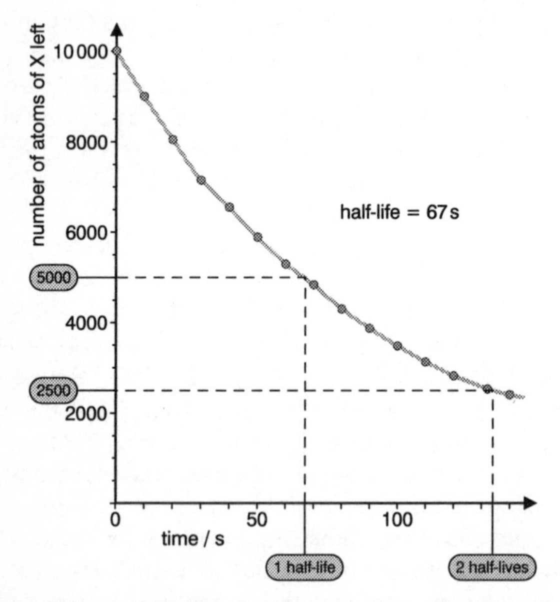

Figure 11D A half life curve

then removed. With such a large number to start with, the number of dice removed would be about 167 (= 1000 / 6) because, on average, one-sixth would show a '1'. If the process was repeated with the 833 remaining dice (= 1000 – 167), the number removed in this second throw would be about 139 (= 833 / 6). If the process were to be repeated a number of times, each time using the remaining dice, the number remaining would decrease as below.,

Number of throws	0	1	2	3	4	5
Number of dice remaining	1000	833	694	578	482	402
Number of dice removed	167	139	116	96	80	67

A graph of the number of dice remaining against the number of throws is the same shape as the half-life curve in Fig. 11D. The number of dice remaining drops to about half in less than four throws. The 'half life' of this process is about four throws therefore. You can prove for yourself that the number would drop to about 250 in about four more throws.

Radioactive hazards

Radioactivity is dangerous to human health because ionizing radiation kills living cells and causes tumours. For this reason, the use of radioactive sources is subject to strict regulations. For example, radioactive sources must be kept in a lead container to prevent radiation from escaping. Also, radioactive sources should only be moved using a suitable handling device.

Because radioactive sources with long half lives remain radioactive for many years, disposal of radioactive sources and products is subjected to legal regulations. Radioactive waste from nuclear reactors must be stored in sealed containers at approved sites.

Radioactive dating enables the age of ancient objects to be measured. For example, living wood contains a small proportion of a radioactive isotope of carbon, ${}^{14}_{6}C$. When a tree dies, the proportion of this isotope gradually decreases as these atoms disintegrate with a half life of about 5600 years. To measure the age of an ancient wooden object, the count rate due to a sample of the object is measured and compared with the count rate of an equal mass of wood from a living tree. This comparison can then be used to work out how many half lives have elapsed since the tree used to make the ancient object died. If the count rate was reduced to one quarter, the object would be 2 × 5600 years old (= 2 half lives).

Summary

Alpha radiation (α) consists of particles, each composed of two protons and two neutrons.

Beta radiation (β) consists of electrons, each emitted when a nucleus with too many neutrons disintegrates.

Gamma radiation (γ) consists of high energy photons.

The half life of a radioactive isotope is the time taken for half the number of atoms of the isotope to disintegrate.

Questions

Q4. The following measurements were made using a geiger counter and a radioactive source with a long half life:

1. Number of counts in 600 seconds with no source present = 245.
2. Number of counts in 100 seconds with the source 20 mm from the geiger tube = 455, 461, 438.
3. Number of counts in 100 seconds with a 1 mm metal plate between the source and the tube = 325, 340, 315.

(a) Calculate the background count rate in counts per second.

(b) Show that the corrected count rate due to the source without the plate present has an average value of 4.10 counts per second.

(c) Show that the corrected count rate due to the source with the plate present has an average value of 2.86 counts per second.

(d) Show that 70% of the radiation incident on the plate passes through it.

(e) What type of radiation is emitted by the source?

Q5. A certain radioactive isotope has a half life of 8 hours. A solution containing 500 million atoms of this isotope is prepared. How many atoms of this isotope have not disintegrated after (a) 8 hours, (b) 24 hours, (c) one week?

Quarks and leptons

Antimatter was predicted by the British physicist Paul Dirac in 1926. He reckoned that for every type of particle, there is a corresponding antiparticle with identical mass and the opposite type of charge if the particle is charged. He predicted that a particle and its antiparticle can be created as a pair and they can annihilate each other. Within a few years, the American physicist Carl Anderson had detected the **positron**, the antiparticle of the electron. Anderson first observed the track of a positron in a device known as a cloud chamber. A charged particle passing through a cloud chamber leaves a track because it creates ions along its path. The ions cause tiny droplets to form from the vapour in the chamber, leaving a visible track. Anderson recognized the track was produced by a particle like the electron but the magnetic field applied to the chamber made the track curve in the opposite direction to that expected for an electron.

Cosmic radiation consists of particles that enter the Earth's atmosphere from space. These particles are produced by the Sun or other stars. They crash into the nuclei of atoms in the atmosphere, creating cascades of particles at high speed that can reach the Earth's surface. As well as positrons, other particles called pi-mesons (or pions) were discovered. These particles were called *mesons* because they were found to be 'middle-weight' particles, heavier than electrons but lighter than protons. Further investigations revealed the existence of 'strange' particles that are created in particle pairs, rather than particle–antiparticle pairs, and which decayed into pions and protons.

Accelerators have been constructed to create and study these new types of particles and antiparticles. In essence, an accelerator is an evacuated tube containing electrodes which are used to accelerate charged particles such as electrons or protons to speeds approaching the speed of light. These charged particles are then directed in a narrow beam at a target. Some of the charged particles collide with the target nuclei to create mesons and other short-lived particles and antiparticles from the energy of the charged particles in the beam. Using these accelerators, physicists have discovered a large number of short-lived particles and antiparticles, charged and uncharged, with a range of masses. These newly discovered particles and antiparticles were found to fit patterns that could be explained by assuming that protons and neutrons are composed of three elementary particles which became known as **quarks**.

The quark model is based on the following assumptions:

1 There are six different types of quarks, the up quark, the down quark, the strange quark, the charmed quark, the bottom quark and the top quark. (referred to as u, d, s, c, b and t for brevity).

2 For every type of quark, there is a corresponding antiquark.

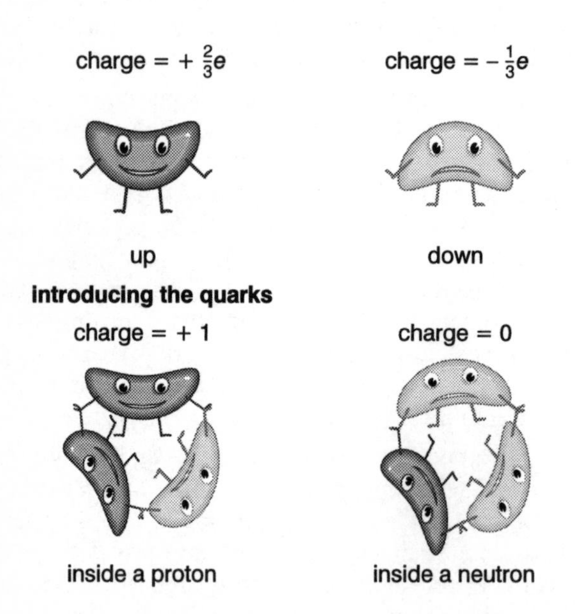

Figure 11E Quarks

3 Quarks combine in threes to form particles like the proton and the neutron. Antiquarks also combine in threes to form antiparticles like the antiproton and the antineutron. Such composite particles are collectively referred to as *baryons*.

4 A meson consists of a quark and an antiquark.

In terms of the charge of the electron, the u, c and t quarks each carry a charge of $+\frac{2}{3}e$ and the other three quarks carry a charge of $-\frac{1}{3}e$. An antiquark carries an equal and opposite charge to its corresponding quark. The symbol for an antiquark is the same as for a quark but with a bar over the top. For example, $\bar{d}$ represents the symbol for a down antiquark.

Thus

- a proton is composed of two up quarks and a down quark,
- a neutron consists of an up quark and two down quarks,
- a pion consists of an up or down quark and an up or down antiquark,
- strange particles contain strange quarks or antiquarks.

The creation of pions and strange particles and antiparticles can be explained using the quark model. For example, if a proton at high speed collides with another proton, the following interaction could take place:

proton + proton → positive pion + proton + neutron

In quark terms,

uud + uud → u$\bar{d}$ + uud + udd

In the collision, a down quark and a down antiquark are created from the kinetic energy of the high speed proton. The quarks and the antiquark regroup to form a positive pion, a proton and a neutron.

The quark model was confirmed by physicists using the Stanford Linear Accelerator to accelerate electrons to speeds within a tiny fraction of the speed of light and use them to bombard a target. These electrons were scattered by the target nuclei in directions corresponding to three hard centres in every neutron and proton.

Where do electrons fit into this model? The answer is that they do not. Electrons, positrons and certain other particles and antiparticles are thought to be elementary in the sense they are not composed of smaller particles. These particles and antiparticles are collectively referred to as **leptons**. All particles not in the quark family belong to the lepton family. Quarks and antiquarks might themselves be composed of even smaller particles. The difference of charge between the up quark and the down quark is equal to the charge of the electron. This perhaps suggests there is a deeper link between quarks and leptons. The Large Hadron Collider at CERN, the European Centre for Nuclear Research, might provide some answers when its construction is completed and it becomes operational. This accelerator is designed to collide protons at energies more than 20 times greater than the biggest accelerators in current use. The mechanism linking mass and energy might at last be uncovered, a century after Einstein discovered the link.

Summary

Quarks

1. There are six different types of quarks, the up quark (u), the down quark (d), the strange quark (s), the charmed quark (c), the bottom quark (b) and the top quark (t).
2. For every type of quark, there is a corresponding antiquark.
3. Quarks combine in threes to form particles like the proton and the neutron, collectively referred to as baryons. Antiquarks also combine in threes to form antibaryons.
4. A meson consists of a quark and an antiquark.
5. The u, c and t quarks each carry a charge of $+{}^{2}/_{3}e$ and the other three quarks carry a charge of $-{}^{1}/_{3}e$.

Leptons

Leptons are electrons, positrons and certain other particles and antiparticles which are thought to be elementary.

Questions

Q6. What is the total charge of a baryon that consists of (a) an up quark, a down quark and a strange quark, (b) three up quarks, (c) three down quarks?

Q7. What is the total charge of a meson that consists of (a) an up antiquark and a down quark, (b) a strange antiquark and a down quark, (c) a down antiquark and an up quark?

Q8. Write down the quark composition of (a) a proton, (b) a neutron.

Q9. When a beta particle is emitted from an unstable nucleus, a neutron in the nucleus changes into a proton. What change in the quark composition of the neutron takes place in this process?

Q10. A proton moving at high speed collides with a neutron to create a strange quark and a strange antiquark. The quarks and the antiquark regroup to form three composite particles which include a strange meson, a strange baryon and a third particle. Identify the quark composition of the third particle.

proton	+	neutron	→	strange meson	+	a strange baryon	+	?
uud		udd		$d\bar{s}$		sdd		

12 ENERGY FROM THE NUCLEUS

No other branch of science has so far had an impact on the human race like nuclear physics has. Maybe this is not surprising as nuclear weapons have the capacity to destroy the human race. Yet before the discovery of nuclear fission, Lord Rutherford reckoned that obtaining energy on a large scale from the nucleus was 'moonshine'. Nuclear fission, the splitting of a large nucleus, was discovered by Otto Hahn and Fritz Strassmann in Berlin in 1938. Physicists around the world recognized the significance of this discovery as the energy released in this process is much greater than any other process. The British and US Governments recognized the dreadful military power the Nazi dictatorship in Germany would possess if Germany became the first nation to possess nuclear weapons. The Manhattan project was set up in America to make a nuclear weapon, the atom bomb, before such a weapon could be made in Germany. In fact, Germany had been defeated when the first atom bombs were dropped on the Japanese cities of Nagasaki and Hiroshima bringing the Second World War to an end. Nuclear reactors for electricity generation were subsequently developed and now account for a quarter of Britain's electricity generation. Concerns about radioactive waste and the Chernobyl disaster have caused nuclear power to fall out of favour but the current generation of nuclear power stations will need to be replaced if electricity cuts and rationing are to be avoided.

In this chapter, we will look at the principles of nuclear fission, how nuclear reactors work and why they are likely to continue to make a siginificant contribution to our electricity supplies.

Nuclear fission

A radioactive substance emits radiation because the nuclei of its atoms are unstable and disintegrate at random, emitting α radiation or β radiation or γ radiation. Such changes causes energy release on a scale about a million times greater than when atoms react chemically. However, radioactive isotopes with sufficiently long half lives are not active enough to release energy at a fast enough rate to produce electricity on a large scale. This is why Rutherford reckoned nuclear power was unrealistic – until **nuclear fission** was discovered.

Nuclear fission is the splitting of a large nucleus into two approximately equal fragments. Energy is released in this process on a scale even greater than when radioactive disintegration occurs. Hahn and Strassmann proved this process happened to the uranium isotope $^{235}_{92}U$. Further investigations showed that this isotope could be made to fission by bombarding it with neutrons and that two or three neutrons are released when such a nucleus splits. These released neutrons are referred to as fission neutrons. Thus a chain reaction is possible in which a neutron splits a uranium 235 nucleus and two or three fission neutrons are released which go on to split other uranium 235 nuclei, leading to more neutrons being released which then go on to split more uranium 235 nuclei. Each fission event releases energy and so an enormous amount of energy is released if the chain reaction is maintained.

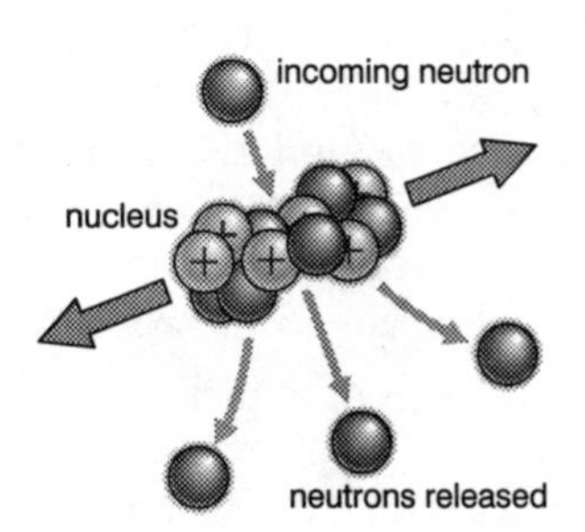

Figure 12A Fission

Binding energy

Why should a large unstable nucleus release energy when it fissions or in a radioactive change? The potential energy of a system depends on the positions of the particles in the system, relative to each other. A stable system is one in which the potential energy of the system is at its lowest. When an unstable system becomes more stable, it changes to a state of lower potential energy. The protons and the neutrons in a nucleus are held together by a strong attractive force that prevents the protons pushing away from one another. To separate the protons and neutrons from one another work would need to be done on them to overcome the strong nuclear force. The work needed to separate a nucleus into separate neutrons and protons is referred to as the **binding energy** of the nucleus. The greater the binding energy of a nucleus, the greater the work that would be needed to separate the neutrons and the protons in the nucleus from each other.

The mass of a nucleus is less than the mass of the same number of separate neutrons and protons. For example, the mass of a helium nucleus which consists of two protons and two neutrons is 0.8% less than the mass of two protons and two neutrons separated from each other. This difference is called the *mass defect* of the nucleus and is due to the protons and neutrons binding together when the nucleus was formed. The binding energy of the nucleus can be calculated from the mass defect using Einstein's famous equation $E = mc^2$.

$$\textit{Binding energy} = \textit{mass defect} \times c^2$$

Nuclear masses are usually expressed in atomic mass units (u) using the carbon 12 scale (see p. 139) and the energy released by a nucleus is usually expressed in millions of **electron volts** (MeV) where 1 MeV = the energy gained by an electron accelerated through a million volts = 1.6×10^{-13}J. Using the above equation, a mass defect of exactly 1 u equates to 931 MeV of binding energy.

The binding energy per nucleon of a nucleus is the binding energy of a nucleus divided by the number of nucleons (i.e. protons and neutrons) in the nucleus. This quantity is a measure of the stability of a nucleus. It can be easily calculated for any nucleus ^A_ZX of known mass M by following the steps below:

1 The mass defect (in atomic mass units) of the nucleus, $\Delta m = Z\,m_p + (A–Z)\,m_N – M$ where m_p is the mass of a proton and m_N is the mass of a neutron.

2 The binding energy E_b (in MeV) = $931 \times \Delta m$.

3 The binding energy per nucleon = E_b / A.

Note: Z = the number of protons in the nucleus, A = the number of neutrons and protons, so A–Z is the number of neutrons in the nucleus.

Worked example

The mass of a $^{235}_{92}U$ nucleus is 234.99333 u. The mass of a proton = 1.00728 u and the mass of a neutron = 1.00867 u. Calculate the binding energy per nucleon of a $^{235}_{92}U$ nucleus.

Solution

$Z = 92$, $A = 235$ $\therefore$ the number of neutrons = $A–Z$ = 143

$\therefore$ Mass defect = $(92 \times 1.00728) + (143 \times 1.00867) - 234.99333 =$ 1.91624 u

$\therefore$ Binding energy = $1.91624 \times 931 = 1784$ MeV

$\therefore$ Binding energy per nucleon = 1784 / 235 = 7.6 MeV per nucleon.

A graph of binding energy per nucleon against nucleon number A is shown in Fig. 12B. Remember the greater the binding energy per nucleon of a nucleus is, the more stable the nucleus is. The graph shows that

1. the binding energy per nucleon increases as A increases to a maximum of about 9 MeV per nucleon at about A = 50 to 60 then decreases gradually,
2. the most stable nuclei are at about A = 50 to 60 since this is where the binding energy per nucleon is greatest,
3. the binding energy per nucleon is increased when nuclear fission of a uranium 235 nucleus occurs,
4. the binding energy per nucleon is increased when light nuclei are fused together.

When a uranium 235 nucleus undergoes fission, the two fragment nuclei each comprise about half the number of nucleons. Therefore the binding energy per nucleon increases from about 7.5 MeV per nucleon for uranium 235 to about 8.5 MeV per nucleon for the fragments. Thus the binding energy per nucleon increases by about 1 MeV for every nucleon which means that the energy released from the fission of a single nucleus is about 200 MeV. The mass of

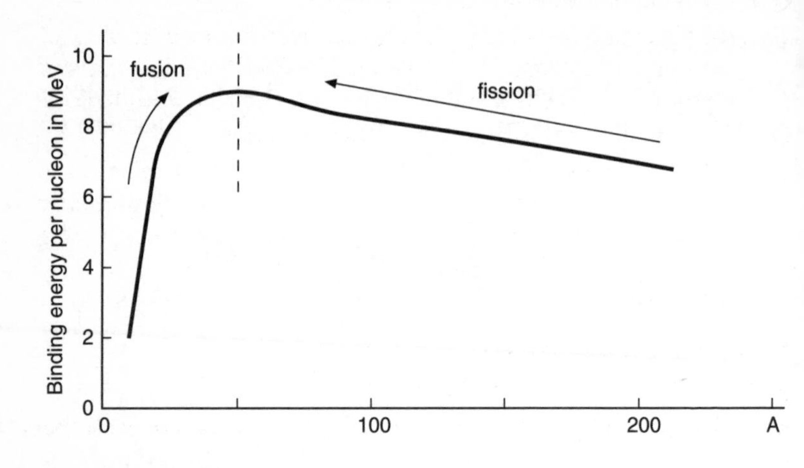

Figure 12B Binding energy

a uranium 235 nucleus is about 4×10^{-25} kg. Prove for yourself that 1 kilogram of uranium 235 would release about 100 million million joules if completely fissioned. In contrast, the energy released by burning fossil fuel is about 10 million joules per kilogram.

Summary

Nuclear fission is the splitting of a large nucleus into two approximately equal fragments.

The binding energy of a nucleus is the work needed to separate a nucleus into separate neutrons and protons.

Binding energy E_b (in MeV) = $931 \times \Delta m$, where Δm is the mass defect (in u) of the nucleus.

Questions

Q1. A neutron strikes a $^{235}_{92}U$ nucleus and causes it to fission. Three neutrons are released in this process. One of the fission fragments is the xenon isotope $^{136}_{54}Xe$. How many neutrons and how many protons does the other fission fragment contain?

Q2. (a) What is meant by binding energy?

(b) Calculate the binding energy per nucleon of a nucleus of the uranium isotope $^{238}_{92}U$ (mass = 238.05076 u). The mass of a proton = 1.00728 u and the mass of a neutron = 1.00867 u.

Q3. (a) Explain what is meant by a chain reaction.

(b) In a certain chain reaction, each fission event releases on average two neutrons that go on to cause further fission after 0.1 seconds. How many nuclei would have been fissioned 1 second after an initial fission event?

Nuclear reactors

A nuclear reactor in normal operation releases energy at a steady rate. This is made possible because the rate of fission events in the reactor is controlled by *control rods* so that each fission event causes exactly one further fission event. The control rods are made of material that absorbs neutrons and they can be moved in or out of the reactor core to keep the number of neutrons in the core constant. Should the fission rate rise, the control rods are pushed into the core a little to mop up surplus neutrons and thus reduce the fission rate.

The mass of uranium 235 must exceed the *critical mass* which is the least mass for fission to be sustained. If the mass of uranium 235 is less than the critical mass, too many neutrons escape or are absorbed by uranium 238 if this is present. The fuel in a nuclear reactor is enriched uranium contained in sealed fuel rods. Natural uranium is mostly uranium 238 ($^{238}_{92}U$) and it contains less than 1% uranium 235 ($^{235}_{92}U$). Uranium 238 absorbs neutrons without fission. For this reason, the percentage of uranium 235 must be increased to about 2–3% (i.e. enriched) otherwise neutrons are absorbed without fission or they escape from the reactor core.

The first atom bomb consisted of two sub-critical hemispheres of pure uranium 235 placed at opposite ends of a hollow cylinder. When the device was detonated, the two hemispheres were forced together into a super-critical mass which then exploded.

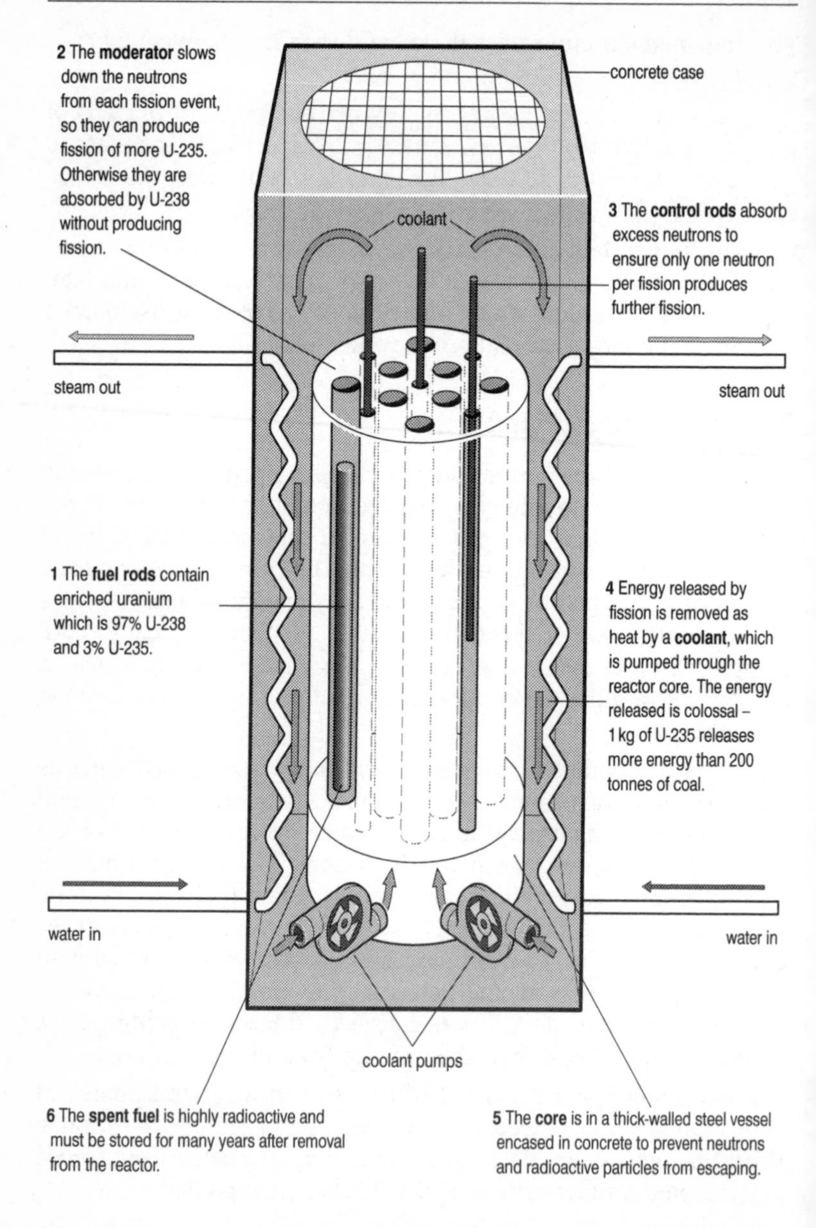

Figure 12C A thermal nuclear reactor

Thermal nuclear reactors

The neutrons released when a uranium 235 nucleus fissions in a nuclear reactor move too fast to cause further fission so they need to be slowed down. This is achieved by surrounding the fuel rods that contain the uranium 235 with a suitable substance, a *moderator*, that 'moderates' the speed of the neutrons. The moderator needs to be composed of light atoms. The neutrons collide with the moderator atoms and transfer kinetic energy in the process to the atoms. This transfer process is most effective if the moderator atoms are as light as possible. The moderator must also be unreactive. For these reasons, water or graphite (which is carbon) is used as the moderator. After many collisions with moderator atoms, the neutrons move slowly enough to cause further fission. They are then referred to as *thermal* neutrons because their kinetic energies are of the same order of magnitude as the moderator atoms.

The design of one type of thermal nuclear reactor is shown in Fig. 12C. The fuel rods are placed in channels in the solid graphite core. The control rods are in channels between the fuel rods. The moderator becomes very hot because its atoms absorb kinetic energy from the neutrons so carbon dioxide gas as a coolant fluid is pumped through the moderator channels. The entire core is enclosed in a sealed steel vessel which is connected via inlet and outlet pipes to a heat exchanger. The coolant is pumped round this sealed circuit and used to raise steam in the heat exchanger to drive electricity generators.

Fuel usage

The complete fission of a kilogram of uranium 235 releases about 8×10^{13}J of energy (= 80 million million joules). A 1000 MW power station operating at an efficiency of 25% needs 4000 million joules of energy from its fuel every second. Each day, the fuel would therefore need to release about 350 million million joules. About 4 kilograms of uranium 235 would therefore be needed to keep a 1000 MW nuclear reactor operating for a day. If the fuel for the reactor contained 2% uranium 235, then the mass of fuel needed per day would be about 200 kg (= 0.2 tonnes). In contrast, a 1000 MW oil-fired power station uses about 40 000 tonnes of oil per day.

Chernobyl

In 1986, a nuclear reactor at Chernobyl in the then USSR exploded and scattered radioactive substances into the atmosphere and the surrounding land. Within days, these substances were scattered by the atmosphere over other countries including Britain. Upland regions in Britain that experienced rainfall in this period were most affected as the rain water contained particles of radioactive isotopes. The fuel rods in a nuclear reactor contain many different radioactive isotopes including uranium 238, plutonium 239 (produced as a result of uranium 238 absorbing neutrons) and neutron-rich fission fragments. The half lives of these isotopes range from fractions of a second to millions of years. Background count rates increased temporarily in the weeks after the Chernobyl disaster. Many thousands of people in the districts surrounding Chernobyl were moved permanently to other parts of the USSR. Children have been most affected as a result of illnesses such as leukemia caused by radioactive particles entering their bodies. The reactor used steam as the coolant. The cause of the disaster seems to have been an interruption in the coolant flow which allowed the core to overheat. The steam reacted chemically with the moderator to produce hydrogen which exploded and blew the top off the reactor vessel. Many of the emergency workers died subsequently from diseases caused by exposure to radioactivity.

Summary

For a steady chain reaction in a nuclear reactor, exactly one fission neutron per fission event must go on to produce a further fission event.

A thermal nuclear reactor contains a moderator which slows fission neutrons down to enable them to produce further fission of uranium 235.

Questions

Q4. Explain what is meant by the term *critical mass* in relation to nuclear fission.

Q5. Explain the action of (a) the moderator, (b) the coolant in a nuclear reactor.

Q6. Use the following data to calculate the mass of fuel used per week in 1000 MW nuclear reactor operating at an efficiency of 25%.

Energy released per kilogram of uranium 235 = 8×10^{13}J; uranium 235 content of the fuel = 2%

Energy options for the future

The demand for energy is likely to increase in the next few decades. As we saw in Chapter 6, oil and gas supply more than 60% of the world's energy supplies. However, present known reserves will probably be used up within the next 50 years. Coal provides about a third of the world's energy supplies at present. Known reserves of coal could last up to three centuries at the present rate of usage. About 8% of worldwide energy demand is met by nuclear power stations as they provide between a quarter and a third of the electricity supplies in industrial countries. World reserves of uranium will probably last about 50 years at the present rate of use. Renewable sources of energy such as hydroelectricity supply less than 3% of the world's energy demand. When the world's oil and gas reserves are used up, coal could be used instead but the increased rate of use would shorten the lifetime of the world's coal reserves to little more than a century. In addition, greenhouse gases produced by burning fossil fuels would still be produced at an excessive rate. The energy options for the future are probably limited to the increased use of nuclear power and the development of renewable energy resources on a much larger scale than at present. We looked at different renewable energy resources in Chapter 6. World uranium reserves are unlikely to last beyond the end of the century and nuclear reactors are not in favour politically at present. In this chapter, we will look at two other nuclear options which could be developed more.

Nuclear fusion

The Sun radiates energy at a colossal rate of 400 million million million million joules every second. A millionth of the energy radiated by the Sun in 1 second would be enough to meet the world's energy needs for a whole year. The Sun produces energy in its core as a result of fusing hydrogen nuclei to form helium nuclei. This process started about 5000 million years ago when the Sun was formed and is likely to continue for another 5000 million years before the Sun runs out of hydrogen in its core.

To make hydrogen atoms fuse to form helium atoms, the atoms of hydrogen must collide at speeds over a thousand times faster than they would have in the Earth's atmosphere. The steps in the process of the formation of a helium nucleus from hydrogen nuclei are as follows:

1. Two hydrogen nuclei colliding at such speeds fuse to form a nucleus of the hydrogen isotope, deuterium ${}^{2}_{1}H$.
2. A high speed collision between a deuterium nucleus and another hydrogen nucleus causes a nucleus of the helium isotope ${}^{3}_{2}He$ to form.
3. A collision between two of these helium nuclei would cause the formation of a nucleus of the stable helium isotope ${}^{4}_{2}He$ and the release of two protons.

The very high speeds necessary for the process to work are achieved in the Sun because the energy released heats its core to more than 10 million degrees. Such high temperature conditions have been achieved briefly in nuclear fusion reactors, enabling light nuclei to be fused and release energy in the process. In such conditions, the atoms lose their electrons and the nuclei move about as a 'plasma' without carrying electrons with them. The plasma is contained in a doughnut-shaped tube by means of magnetic fields but it loses all its energy if it touches the tube. This containment problem is the reason why fusion at present cannot be made to last more than a short time. Thus nuclear fusion reactors remain under development at present and are unlikely to contribute significantly to world energy supplies for at least another 50 years.

Fast breeder reactors

Natural uranium contains less than 1% of uranium 235. The other 99% is uranium 238. The spent fuel rods from a thermal nuclear reactor contain unused uranium 238 and the plutonium isotope ${}^{239}_{94}Pu$. Plutonium is an artificial element that is made in a nuclear reactor when uranium 238 nuclei absorb neutrons. Plutonium 239 can be fissioned by fast neutrons. Thus a chain reaction is possible without the need for a moderator. A reactor that uses plutonium 239 as its fuel is called a *fast breeder* reactor because

- the fuel is fissioned by fast neutrons, and
- the reactor can breed its own fuel by allowing neutrons from the plutonium core to be absorbed by a 'blanket' of uranium 238 surrounding the core.

Thus a fast breeder reactor creates its own fuel in the form of plutonium 239 from uranium 238. The entire content of natural uranium can thus be fissioned, not just the uranium 235 content.

Fast breeder reactors would therefore extend the lifetime of the world's uranium supplies from a century or so to several thousand years. However, the spent fuel would still contain many radioactive isotopes from the fission fragments and the danger of plutonium leakage to the environment would increase as greater quantities of plutonium would be in use. Plutonium 239 has a half life of 24 000 years and is much more radioactive than uranium 238 or uranium 235. A further danger is that plutonium 239 could accumulate inadvertently in the process of manufacturing the fuel rods; if too much plutonium accumulated, an uncontrollable chain reaction could be started, ending in a nuclear explosion. The experimental fast breeder reactor at Dounreay in Scotland proved that such a reactor works but it has been decommissioned because of the risk of plutonium leakage. New fast breeder reactors are unlikely to be constructed in the foreseeable future. However, unless renewable resources are developed on a much larger scale, plutonium reactors may be necessary.

Summary

Nuclear fusion is the process of fusing light nuclei together to form heavier nuclei. Nuclear fusion reactors are still in the developmental stage. Fast breeder reactors using plutonium could extend the lifetime of the world's uranium reserves by hundreds of years. Renewable energy resources need to be developed on a much larger scale than at present if many more nuclear reactors are not built.

Questions

Q7. State one advantage and one disadvantage of (a) coal and (b) uranium as a fuel to provide electricity.

Q8. (a) What is meant by nuclear fusion?

(b) Give one technical reason why nuclear fusion is difficult to sustain in a reactor.

Q9. (a) What type of fuel is used in a fast breeder reactor?

(b) Give one advantage and one disadvantage of the fast breeder reactor compared with the thermal nuclear reactor.

Q10. Discuss the range of energy options (including renewable resources described in Chapter 6) that are likely to be available after the world's oil and gas reserves are used up.

13 JOURNEY INTO SPACE

One of the great successes of modern science is the discovery that the Universe originated about 12 000 million years ago in a cataclysmic explosion, the Big Bang, which created space and time. The enormous energy released in the Big Bang created particles and antiparticles which annihilated to produce radiation and excess particles. The entire Universe has been expanding ever since the Big Bang. All the matter in the Universe has formed from the excess particles left after the early stages of the Big Bang. How will the Universe change in future? This is a problem that physicists may well be able to answer soon. In this chapter, we will look at the discoveries that have led to our present state of knowledge.

Physics and astronomy have always been at the cutting edge of human endeavour. Newton put the two subjects on a firm mathematical basis when he wrote the Principia Mathematica, published in 1687, in which he set out the mathematical principles of physics. He showed that these principles govern the motion of objects in space as well as objects on the Earth. Newton's laws of motion and of gravity provided the framework for the development of physics for over two centuries until Einstein produced his theories of relativity. We will look in this chapter at Newton's theory of gravity and Einstein's theory of general relativity which replaced it and which led to predictions now confirmed such as the bending of light by gravity, black holes and the expansion of the Universe.

About gravity

From Earth, the planets appear as wandering stars because each planet, as seen against the distant stars, changes its position relative to the Earth as it orbits the Sun. Johann Kepler (1571–1630) devised three laws to describe the motion of the planets, using his own observations and those of his predecessor in Prague, Tycho Brahe (1546–1601), the foremost astronomer of his era.

Kepler's 1st Law: each planet moves on an elliptical path round the Sun.

Kepler's 2nd Law: the straight line from the Sun to a planet sweeps out equal areas in equal times as the planet moves round the Sun.

Kepler's 3rd Law: the cube of the mean radius of the orbit of a planet is in proportion to the square of the time taken by the planet for each orbit.

- Kepler's 1st law introduced the new idea that the orbits of the planets are in general ellipses, not circles. Figure 13A shows how to sketch an ellipse. Kepler said that each planet's orbit was an ellipse with the Sun at one of the focal points of the ellipse.
- Kepler's 2nd law arose because he knew from his observations that the progress of Mars at its maximum

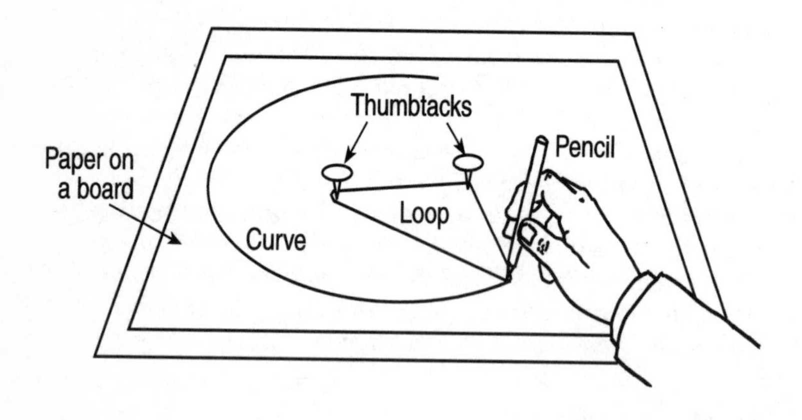

Figure 13A (a) Making an ellipse

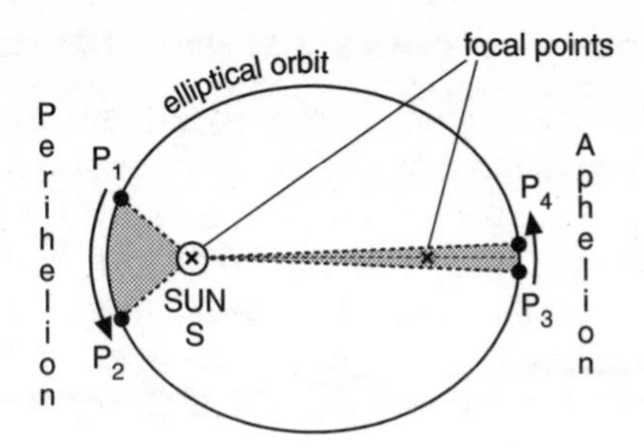

Time from $P_1 \rightarrow P_2$ = Time from P_3 to P_4
if area SP_1P_2 = area SP_3P_4

(b) Kepler's 2nd Law

distance from the Sun is 0.8 times slower than its progress at its least distance. He thus worked out that if an imaginary straight line from the Sun to Mars took a certain amount of time to sweep through an angle of 10° at minimum distance, the line would sweep through an angle of 8° in the same time at maximum distance. He also knew that the least distance from the Sun to the planet Mars is about 0.9 times the maximum distance and he showed that a sweep of 8° at maximum distance covers an area equal to the area swept out for 10° at minimum distance.

- Kepler's 3rd law required the measurement of the mean radius of orbit, *r*, of each planet and the time, *T*, each planet takes to orbit the Sun. The results of these measurements are given in the table below. Note that the value of r^3 / T^2 is the same for every planet. Thus Kepler's 3rd Law follows, namely r^3 is proportional to T^2.

	Mercury	**Venus**	**Earth**	**Mars**	**Jupiter**	**Saturn**
r in A.U.	0.39	0.72	1.0	1.5	5.2	9.5
T in years	0.24	0.61	1.0	1.9	11.9	29.5
r^3 / T^2	1.0	1.0	1.0	1.0	1.0	1.0

1 A.U. = 1 astronomical unit = the mean distance from the Sun to the Earth.

Newton's theory of universal gravitation

The reason why the planets move round the Sun remained a mystery until Newton developed the theory of gravity. Newton had already worked out the laws of motion (see pp. 37–41) and he knew that a force must act on each planet to keep it moving round the Sun. He realized that this force must be an attractive force and that it depended on the mass of the planet, the mass of the Sun and the distance from the planet to the Sun. He then deduced that the force of gravity varies with the inverse of the square of the distance from the planet to the centre of the Sun. In other words, at 2 AU from the Sun, the force of gravity on an object due to the Sun is a quarter what it would be on the same object at 1 AU; at 3 AU, the force would be a ninth what it would be on the same object at 1 AU.

Newton made this prediction by taking the steps below:

- He worked out the theory of motion for any object moving round a circle of radius r at a constant speed υ and proved that the force needed is proportional to υ^2 / r.

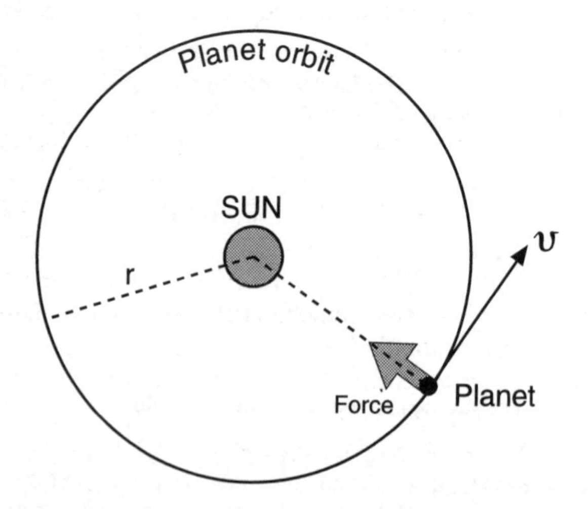

Figure 13B Newton's theory of universal gravitation

- From Kepler's 3rd Law, he knew that r^3 / T^2 is constant and as the speed υ is equal to $2\pi r / T$ where $2\pi r$ is the circumference of the orbit, he deduced that $\upsilon^2 r$ is constant.
- Newton then showed that as $\upsilon^2 r$ is constant, then υ^2 / r must be proportional to $1 / r^2$.

Thus the force of gravitational attraction between the Sun and a planet is proportional to $1 / r^2$.

Newton put forward the theory that the force of gravitational attraction between any two objects is

1. proportional to the product of the masses of the objects,
2. inversely proportional to the square of the distance between the centres of the objects.

He then proved the general result that this force law caused planets to move on elliptical paths round the Sun. He also showed that the inverse square law explained the motion of comets round the Sun, the Moon's orbital motion round the Earth and how the Moon's gravity causes the ocean tides on the Earth. Newton's theory of gravity is described as a universal theory because it explains the motion of objects on the Earth as well as in space. However, a minor problem was discovered in the nineteenth century in connection with the orbit of Mercury that could not be explained by Newton's theory. The problem remained until Einstein solved it using his general theory of relativity, a theory which drastically altered our understanding of space and time. We shall look at the consequences of this revolutionary theory in the next part of this chapter.

How to lose weight

The pull of gravity on an object on the Moon is one-sixth of the pull of gravity on the same object on the Earth. A 60 kilogram person has a weight of about 600 newtons on the Earth but only about 100 newtons on the Moon.

- The Moon's mass is about 1% of the Earth's mass which would reduce lunar gravity to 1% of Earth's gravity if the moon had the same radius as the Earth.
- However, the Moon is one-quarter the size of the Earth so its surface gravity is 16 times stronger

than if the Earth and Moon were the same size.

- Hence lunar gravity is 16% of Earth's gravity (which is about one-sixth).

Einstein's new universe

In his 1905 theory of relativity, Einstein showed that space and time are not separate quantities. By assuming that the speed of light is independent of the motion of the source and of any observer, he predicted such effects as length contraction, time dilation and relativitistic mass. See p. 167. In his theory, he showed that all uniform motion is relative and that absolute uniform motion cannot be detected and is therefore meaningless. In other words, he showed that the laws of physics are the same for all observers moving at constant velocity relative to each other. From 1905, he spent more than ten years developing his ideas on space and time in order to prove the general principle of relativity that the laws of physics are the same for all observers in non-uniform motion as well as in uniform motion. In the course of this work, he proved that accelerated motion and motion due to gravity cannot be distinguished and are therefore equivalent. Thus he reduced gravity to a feature of space and time and went on to prove that gravity bends light as a result. He published his general theory of relativity in 1916 and used it to explain the astronomical puzzle about the elliptical orbit of the planet Mercury that had confounded scientists ever since the problem was discovered in the previous century. Careful measurements on the position of the perihelion had revealed that this position was slowly advancing at a rate of 1.2 hundredths of a degree per century. The perihelion is the nearest point along the orbit to the Sun. Einstein proved that this perihelion advance was caused by the Sun's gravity distorting space near to the Sun.

Einstein also predicted that starlight grazing the Sun would be deflected, an effect he said could be measured by observing stars near the Sun during a solar eclipse. When this prediction was confirmed by British astronomers in 1919, Einstein became a worldwide celebrity as 'the person who knew how to bend light'. More significantly, his theory of general relativity was confirmed and he thus proved that absolute space and absolute time do not exist.

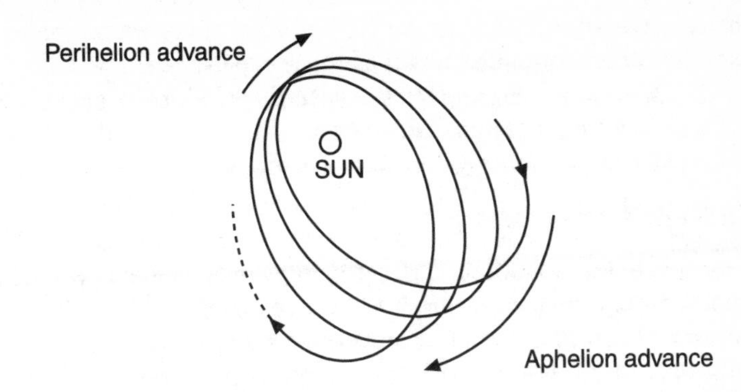

Figure 13C Mercury's orbit

Gravitational lenses and black holes

Further evidence that gravity bends light has been obtained by astronomers in recent years as a result of the observation of distorted images of some distant galaxies. A galaxy is an enormous collection of stars held together by their own gravity. The Sun is part of the Milky Way galaxy which is reckoned to contain more than a thousand million stars. The space between galaxies is thought to be mostly empty. Distorted images are caused by light from a distant galaxy skimming the edge of an invisible dark galaxy before reaching us. The dark galaxy acts as a *gravitational lens* and bends this light, causing us to see a distorted image of the distant galaxy that emitted the light. Often, the galaxy that emitted the light is hidden completely behind the dark galaxy so we see its image only as patchy rings of light. The same effect can be observed if a torch bulb is observed through a glass lens which is partly covered by an opaque disc; the light from the torch bulb reaches your eye via the rim of the lens so you see the rim as a ring of light.

Gravitational lensing provides unexpected evidence in support of Einstein's general theory of relativity. Recent discoveries of *black holes* in space is further evidence in support of Einstein's ideas. Such objects were predicted by Karl Schwarzchild using the general theory of relativity a few years after Einstein published the theory. Not even light can escape from a black hole. Any object inside a black hole cannot escape from it. The mass of a black hole is so large that nothing, not even light, can escape from its gravity. Imagine a rocket projected into space from the surface of a planet.

The rocket will fall back to the surface if its speed of projection is less than a certain value known as the escape speed. The larger the mass of the planet, the bigger the speed needed to escape from it. Nothing can travel faster than light so if the mass of the planet is sufficiently large, nothing can escape from it.

An object that falls into a black hole is trapped forever. The *event horizon* of a black hole is the boundary of a black hole as any object on or inside the boundary is trapped forever. A rocket approaching a black hole could pass through the event horizon into the black hole, thus becoming trapped forever. No signals could be sent from the rocket to observers outside the event horizon because nothing, not even light, can escape from the black hole. Schwarzchild showed that the event horizon of a black hole is a sphere of radius in proportion to its mass. For a black hole of mass equal to the Sun's mass, the radius of the event horizon would be about three kilometres. Such a black hole would pull in nearby stars, becoming even larger in the process.

Evidence for black holes

Astronomers reckon that the galaxy M87 has a black hole at its centre. There could even be a black hole at the centre of the Milky Way, our home galaxy. Fortunately, the Sun is in the outer spiral arms of the galaxy, not at its centre. A star about 10 or more times larger than the Sun would become a black hole if it collapsed without loss of mass to a ball a hundred thousand times smaller. In fact, such a dramatic collapse, known as a *supernova*, happens at the end of the lifecycle of a giant star. The core of such a star collapses when all its light elements have been fused to form heavier elements (see p. 198). The outer layers of the star are thrown off in a massive explosion known as a supernova and the core is compressed so much that a black hole is formed. Such a dramatic end will not happen to the Sun as it is not massive enough.

Summary

The force of gravitational attraction between any two objects is inversely proportional to the square of the distance between the centres of the objects.

Accelerated motion and gravity cannot be distinguished and are therefore equivalent.

Light follows a curved path in a sufficiently strong gravitational field.

The event horizon of a black hole is the boundary of a black hole as any object on or inside the boundary is trapped forever.

Questions

Q1. A comet moves on an elliptical path round the Sun, stretching far beyond the Earth's orbit.

(a) Sketch a diagram to show the orbit of a comet. On your diagram, mark the position of the Sun and draw a circle to represent the Earth's orbit.

(b) Explain why a comet takes far more time to return to the inner solar system than it takes to pass through the inner solar system.

Q2. (a) Jupiter orbits the Sun once every 11.8 years. Use Kepler's 3rd Law to prove that its radius of orbit is 5.2 A.U.

(b) The mean radius of Saturn's orbit is 9.5 A.U. Use Kepler's 3rd Law to prove that Saturn orbits the Sun once every 29 years.

Q3. (a) State two reasons why the Moon's surface gravity is weaker than that of the Earth.

(b) The mass of Mars is 11% of the mass of the Earth. The radius of Mars is 53% of the radius of the Earth. Use this information to show that the surface gravity on Mars is 38% of that of the Earth.

Q4. (a) On a sketch diagram showing the Earth and the Sun, sketch the path of a radio signal sent from a spacecraft to the Earth when their positions are such that the signal grazes the Sun.

(b) With the aid of a sketch diagram, describe how a galaxy can act as a gravitational lens.

Q5. (a) State one characteristic feature of a black hole.

(b) Describe what is meant by the *event horizon* of a black hole.

The expanding universe

By the end of the twentieth century, astronomers had worked out that the Universe is expanding. The distant galaxies are moving away from us as the galaxies move apart due to the expansion of the Universe. However, two centuries earlier, most scientists believed the Universe to be infinite in space and in time, the same now as it always has been. This view was challenged in 1826 by Heinrich Olbers who proved mathematically that the sky would be permanently bright if there was an infinite number of stars. So Olbers concluded that the Universe is finite and he reckoned it must be expanding, otherwise gravitational attraction would cause it to collapse. A century later, Einstein used his general theory of relativity to show that a static finite Universe is possible but he needed to include a cosmic repulsion force to stop the Universe collapsing. The question of whether the Universe is static or in a state of collapse or expansion was settled by the astronomer Edwin Hubble in 1929 when he published the results of his survey of about two dozen galaxies at known distances within about 6 million light years of the Milky Way galaxy. Hubble discovered that the spectrum of light from each galaxy is shifted towards the red part of the spectrum. This shift is known as a **red shift** and it happens because the light waves from a receding light source are lengthened due to the source moving away as the light is being emitted from the source.

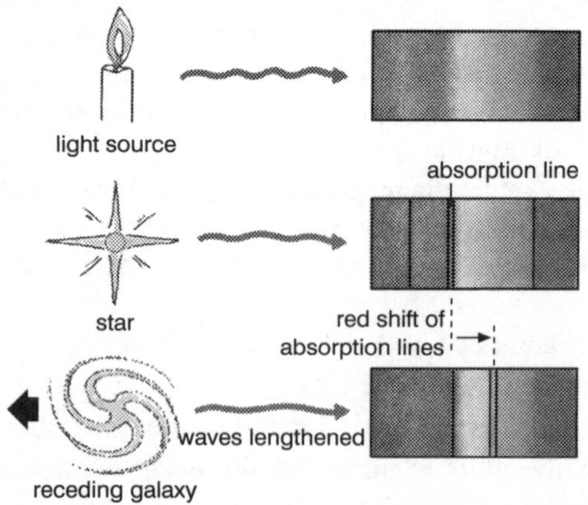

Figure 13D Red shift

Hubble knew that the increase of the wavelength of light due to a red shift is in proportion to the speed at which the light source moves away (i.e. recedes). He measured the wavelength change for the light from each galaxy and used the measurement to calculate the speed of recession of each galaxy. His results showed that the speed of recession of a galaxy is in proportion to the distance from the Sun to the galaxy. This discovery is known as *Hubble's Law*.

Why should all these galaxies be rushing away from us? The Earth has no special place in the Universe. Hubble realized that the Universe is expanding so that

- every galaxy is moving away from every other galaxy,
- the further a galaxy is from us, the greater its speed of recession.

Make a model of the expanding Universe

Mark a rubber band with spots along its length at equal intervals. The spots represent galaxies. Observe how the spots move away from each other when the band is stretched slowly. Two spots initially near to each other move away from each other much slower than two spots initially far from each other.

The Big Bang

If the Universe is expanding, could it have once been much smaller? In fact, the Universe originated in a massive primordial explosion, known as the *Big Bang*, and has been expanding ever since. The Big Bang theory was opposed by some scientists who believed that the Universe is in a steady state, expanding because matter is pouring into it at points in space referred to as 'white holes'. In 1965, the dispute was settled unambiguously by the discovery of microwave radiation from all directions in space. This cosmic microwave background radiation is radiation that has been travelling though space ever since the Big Bang.

More accurate measurements of the red shift and distances to galaxies further and further away continue to support Hubble's Law. These measurements indicate that the speed of a galaxy is 20 kilometres per second for every million light years. Thus a galaxy which is moving away from us at a speed of 20000 kilometres per second must be 1000 million light years distant. Since no object can travel faster than the speed of light which is 300 000 kilometres per second, then the most distant galaxies can not be further away than about 15 000 million light years. Thus the Big Bang cannot have taken place more than 15 000 million years ago which is therefore the age of the Universe. Taking account of gravity, because it holds back the expansion of the Universe, gives an age of about 12 000 million years. In the next section, we will meet quarks and antiquarks again when we consider the chain of events in the early stages of the Big Bang.

The future of the Universe

The theory of the expanding Universe without a cosmic repulsion force had been worked out by Alexander Friedmann in Russia in 1922 using Einstein's general theory of relativity. Einstein described his earlier idea of a cosmic repulsion force as his 'greatest blunder'. However, recent observations indicate that the expansion is accelerating which suggests a force of repulsion could be driving the galaxies away from each other faster and faster.

Einstein's general theory of relativity predicts that the future of the Universe depends on the average density of matter in it which is at present not known.

- If the density is too great, the Universe will stop expanding and then start to contract, ending in the *Big Crunch.*
- If the density is too small, the Universe will expand forever, a scenario known as the *Big Yawn*.
- If the density is equal to a certain value, the Universe will expand at an ever-decreasing rate without ever reversing.

Recent evidence from more accurate measurements on the cosmic microwave background suggests the last scenario. However, the amount of known matter in the Universe is insufficient to prevent a runaway expansion of the Universe. Some form of invisible *dark energy* not yet known about might be making the galaxies accelerate as they move away from each other.

The Early Universe

The cause of the Big Bang is not yet known. What is known is that a vast amount of energy was suddenly released, creating space and time in the process and causing space to expand from nothing to the size of a football in less than a billionth of a billionth of a billionth of a microsecond.

- By this stage, the temperature had fallen to about a million million degrees, allowing quarks, antiquarks, electrons and positrons to form from radiation.
- After about a minute, the temperature had fallen to about a thousand million degrees. The antiquarks were annihilated by the quarks and the positrons were annihilated by the electrons. The remaining quarks formed neutrons and protons. At this stage, the Universe had expanded to a size bigger than the Solar System.
- The neutrons and protons joined together to form nuclei as the Universe cooled further. After about 100000 years, the temperature had dropped to a few thousand degrees. Atoms formed as nuclei were able to hold on to electrons. The Universe became transparent at this stage because atoms absorb light far less than ions do. Thus the dark age of the Universe began.
- After a few billion years, galaxies began to form and move away from each other as the Universe continued to expand. Astronomers do not yet know if stars gathered to form galaxies or if stars formed inside galaxies.

Thus the Universe, which is matter on the largest possible scale, has developed as a result of the properties of matter and radiation on the smallest possible scale. Many questions remain to be answered, including how gravity works and why matter and antimatter did not annihilate each other completely in the early stages of the Big Bang. Clearly, this did not happen otherwise we would not exist.

Summary

The expansion of the Universe: every distant galaxy is moving away from us. The further a galaxy is from us, the greater its speed of recession.

The Big Bang: the Universe originated in a massive primordial explosion, known as the Big Bang, and has been expanding ever since. Quarks, antiquarks, electrons and positrons formed from radiation in the early stages of the Big Bang.

Questions

Q6. (a) Explain what is meant by the 'red shift'.

(b) What did Hubble discover about the link between the red shift and the distance to a galaxy?

(c) What conclusion about the Universe was drawn from Hubble's discovery?

Q7. A galaxy at a distance of 100 million light years was found to be moving away from us at a speed of 2000 kilometres per second. Use this information to estimate

(a) the distance to a galaxy moving away from us at a speed of 8000 km/s,

(b) the speed of recession of a galaxy at a distance of 4000 million light years.

Q8. (a) What is the Big Bang theory of the Universe?

(b) What discovery caused the Steady State theory of the Universe to be rejected in favour of the Big Bang theory?

Q9. Put the following events in order of time of formation, commencing with the Big Bang

Big Bang

Atoms formed

Galaxies formed

Neutrons and protons formed

Quarks and antiquarks formed

Q10. (a) What was the approximate age of the Universe when (i) neutrons and protons formed, (ii) it became transparent?

(b) What is thought to be the most likely future of the Universe?

14 A CHALLENGING FUTURE

In the final chapter, we will look at some of the problems that are at the frontiers of physics at present. The topics chosen are of necessity described briefly and other topics could have been chosen instead. Nevertheless, the aim here is to outline some of the problems of physics at present and to consider the immense benefits that could follow if they are solved. Of course, no one can predict how research into these problems may or may not develop or where new problems and opportunities might arise. At the end of the nineteenth century, many physicists reckoned that little else remained to be discovered except some apparently minor problems to do with radioactivity and light. Ten years into the twentieth century, physics had been revolutionized by the quantum theory and the theory of relativity. In the last few decades, more scientific knowledge has been discovered than in the entire previous history of the human race. Developments in physics in the next few decades may or may not be as revolutionary as a century ago but the work of physicists will undoubtedly continue to provide major benefits to us all.

The frontiers of physics

Understanding energy

We talk about energy, we measure it, we have government departments to control it yet we still do not fully understand it. We can not see it yet we know when an object has energy because we can then make it do some work. Einstein worked out that the mass of an object increases as its speed increases, as outlined on p. 168.

In effect, when the speed of an object approaches the speed of light, the mass of the object increases and its motion becomes more difficult to change. Einstein's equation $E = mc^2$ tells us how much the mass changes for a certain amount of energy transferred. How energy transfer causes a change of mass is not yet known even though Einstein worked out the scale of transfer a century ago. Experiments in high energy physics at laboratories such as CERN, the European Centre for Nuclear Research at Geneva, might provide the answer in the next few years. These experiments are designed to find out why particles such as electrons, protons and neutrons possess mass. An answer to this question could then lead on to finding out how particles with mass gain mass when made to travel faster. The mechanism of $E = mc^2$ would then be uncovered, providing a deeper understanding of nuclear fission and fusion.

Energy for everyone

We all use energy from fuels, especially people in wealthy countries. The rate of use of global energy is about 10 million million watts. Oil, gas and uranium reserves are unlikely to last into the twenty-second century. Coal reserves might perhaps last several centuries. If everyone on the planet were able to use energy at the same rate as in wealthy countries, the rate of global usage would be five times greater. World population is rising fast as well so fuel reserves are likely to become scarcer and more expensive. Wars and conflicts could follow if this happens. New discoveries of fuel reserves will undoubtedly be made but such discoveries are unlikely to bridge the gap between growing demand and declining reserves.

In Chapter 12, we looked at the options for the future in terms of more renewable energy sources and more nuclear power plants. Renewable energy sources using current technology could make a significant contribution but not enough to plug the 'demand gap' at present levels of investment. If energy prices rise, more investment in renewable energy sources is likely but the growth of living standards of the past century might slow down or stop or even reverse if investment is too slow. New discoveries in science could close the demand gap. For example, hydrogen gas could be used as the fuel for vehicles; hydrogen could be produced and collected from sea water by electrolysis. Solar cells could provide the electricity for this process. At present, solar cells are expensive and inefficient. Research into better and cheaper solar cells could result in the 'hydrogen' economy replacing the 'petrol' economy. In addition, such research could produce high voltage solar cells capable of producing much more power and voltage than the present generation of solar cells. Each square metre of a solar panel facing the Sun could absorb as much as 1400 watts of solar power. Even on a dull day, solar radiation could generate 100 watts of electrical power per square metre. To meet global energy demand, less than 0.02% of the surface of the planet (i.e. a total area less than the Sahara desert) would need to be covered in solar panels. No nuclear waste, no greenhouse gases, no oil pollution, no acid rain!

New materials

The benefits of new materials reach us all in many ways, for example through lighter vehicles, better thermal insulation, more reliable artificial joints and improved communications. The explosive growth of the internet is the result of the invention of the transistor in the mid-twentieth century and the subsequent development of integrated circuits on silicon chips in the 1970s. Almost everything we do has been affected by internet and multimedia technologies. Further developments in communications such as cable TV have taken place as the result of research into flexible transparent materials used to make optical fibres. Such fibres are capable of carrying much more information than copper wires carry.

Another potential revolution awaits the discovery of room temperature superconductors. At present, superconductors, which are electrical conductors with zero resistance, need to be cooled to very low temperatures. As the temperature of a superconductor is raised, its resistance suddenly returns at above a certain temperature referred to as its critical temperature. The highest known critical temperature is not much more than about 130 K, well below room temperature which is about 290 K. The benefits of room temperature superconductors would include much more efficient transmission of electrical power across long distances, lighter electric motors, economic magnetic levitation transport systems, cheaper medical scanners and faster computers. Superconducting cables could be used to transfer electrical power long distances from solar power stations in remote areas to populated areas. Research into superconductivity continues as scientists attempt to find out if room temperature superconductivity is possible.

The World Wide Web

The Internet is a communications network that was devised by the US government in the Cold War to ensure an enemy country could not knock out vital communications by means of a nuclear strike. A network of powerful computers in different locations were linked to each other to provide many routes between any two computers. An internet message from one computer to another is chopped into a sequence of packets and each packet is sent independently to the same destination by any route. At the destination, the packets are put together to recreate the original information. The individual packets travel via different routes to the same destination. The Internet now consists of a permanent network of internet service providers (ISPs) connected to the telephone system. The World Wide Web (WWW) was developed in the 1990s by physicists at CERN to enable physicists in other countries to gain access to scientific papers stored electronically via the Internet. Within a few years, WWW revolutionized commerce, industry, entertainment and communications. Few people recognized the immense potential of the World Wide Web at the time it was set up.

An unpredictable future

Blue skies research

No one can tell the effects of a new discovery. Michael Faraday could not have predicted that his discovery of electromagnetic induction (see p. 104) would develop into a worldwide industry and provide energy to consumers on demand. Joseph Thomson, the discoverer of the electron (see p. 174), saw within his own lifetime the development of the electronics industry but even he would have been amazed at the wide range of electronics applications now. Lord Rutherford's investigations on the structure of the atom (see p. 176) led to the discovery of the nucleus yet even Rutherford at that time could not foresee nuclear power and the development of the nuclear industry. Scientific research that might seem obscure or without immediate goals, often referred to as 'blue skies research', has on many occasions produced revolutionary discoveries.

Scientific research driven by precise aims and targets is necessary in companies with expensive equipment where investors expect dividends. Such applied research often leads to important discoveries and developments such as integrated circuits, providing huge returns for investors and inventors. However, blue skies research can revolutionize our understanding of nature as well as developing the essential underpinning knowledge from which revolutionary inventions and developments often follow. A flavour of such research in two areas of physics is outlined below.

Low temperature physics is a very active research field as scientists attain lower and lower temperatures no more than a fraction of a degree above the absolute zero of temperature. The properties of materials at such very low temperatures provide a rich area of research. **Superconductors** were discovered as a result of measuring the electrical resistance of different conductors at such low temperatures. **Superfluidity**, the complete absence of viscosity in a fluid, was discovered when liquid helium was observed emptying itself out of its glass container. Further investigations showed that this fluid lost its viscosity (i.e. its resistance to flow) suddenly at 2.2 K when it is cooled from above to below this temperature. It also became more than a million times more effective as a heat conductor below this temperature. Since this discovery, superfluidity has also been discovered in a few other fluids. The explanation of superfluidity lies in the sudden change of behaviour of the helium atoms when cooled below 2.2 K. Instead of moving about at random as they do above 2.2 K, the atoms link together in a single quantum state at or below 2.2 K. Superfluidity is an example of quantum behaviour on a scale large enough to see directly thus it provides a deeper understanding of quantum theory. In addition, understanding superfluidity has helped scientists understand superconductivity in metals.

Chaos theory is a recent arrival in science and has opened up many new and interesting applications. The ideas of chaos theory have been applied in such diverse fields as fluid flow, population dynamics in biology, medical research in connection with heart attacks, share prices on the finance markets and asteroid impacts. Chaos theory stems from computer modelling of non-linear systems. Some measurable quantity (or quantities) is used to specify the state of the system at regular intervals. If the value of the quantity at some interval does not maintain a constant relationship to the value in the previous interval, the system is described as *non-linear*. A non-linear system can behave chaotically in certain circumstances whereas a *linear* system where the proportionality is maintained is reliable and predictable.

A simple example of a non-linear system is the motion of an object confined to move along a line such that its position changes according to how far it is from each of two fixed points along the line. Its motion depends on its initial position and can change very dramatically from confined chaos to running away with just a small change of its initial position, as shown in Fig. 14A. The spreadsheet used to generate this model is given in Appendix 1.

To illustrate how the ideas can be applied more widely, consider the population dynamics of animals on a small island. Suppose the island is populated only by a certain animal species which breeds once each year and which lives off the natural vegetation on the island. The number of animals each year depends on the number the previous year. Too many animals would reduce the food supply and cut the number of animals that survive; too few animals would breed too little. Suppose the island can support a maximum of 100 animals each year. The population change each year is proportional to the population P in the previous year due to breeding and to $(100 - P)$ on account of the food supply. In other words, the population in a given year depends on how far the previous year's population was from 0 and 100. The mathematics works out similar to the previous situation except extinction would occur if P reached 100 in any year as there would be no food supply left.

Chaos and catastrophe

The human race on planet Earth is not unlike an animal population on an island. Catastrophe could occur through an asteroid impact

Parameter= 1.1
Initial pos= 0.8

Time	Old position	New position
0	0.8	0.704
1	0.704	0.91689
2	0.91689	0.335292
3	0.335292	0.980633
4	0.980633	0.083564
5	0.083564	0.336955
6	0.336955	0.983032
7	0.983032	0.073394
8	0.073394	0.299231
9	0.299231	0.922645
10	0.922645	0.314035
11	0.314035	0.947834
12	0.947834	0.217556
13	0.217556	0.748991
14	0.748991	0.827215
15	0.827215	0.628894

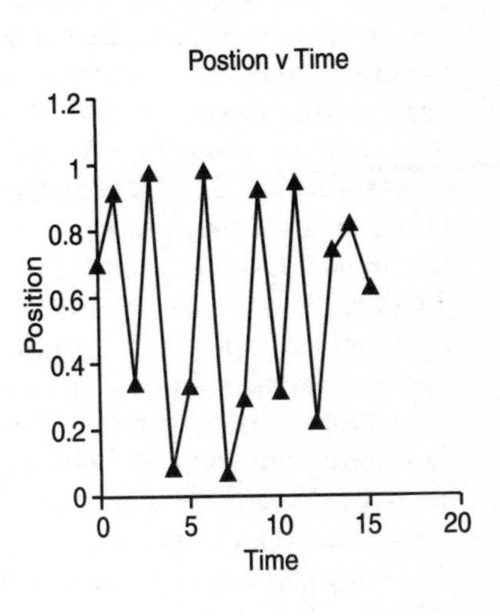

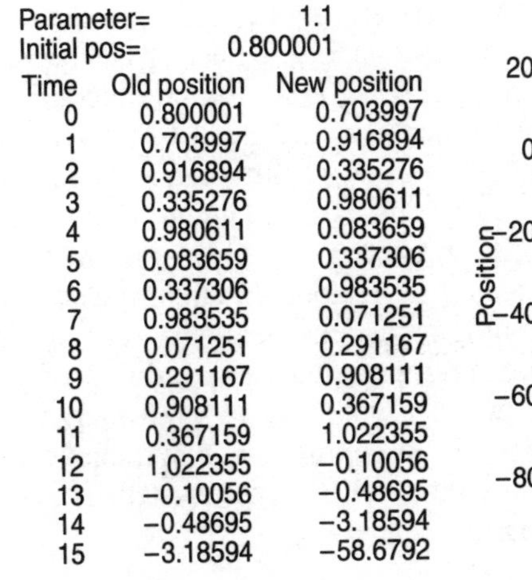

Parameter= 1.1
Initial pos= 0.800001

Time	Old position	New position
0	0.800001	0.703997
1	0.703997	0.916894
2	0.916894	0.335276
3	0.335276	0.980611
4	0.980611	0.083659
5	0.083659	0.337306
6	0.337306	0.983535
7	0.983535	0.071251
8	0.071251	0.291167
9	0.291167	0.908111
10	0.908111	0.367159
11	0.367159	1.022355
12	1.022355	−0.10056
13	−0.10056	−0.48695
14	−0.48695	−3.18594
15	−3.18594	−58.6792

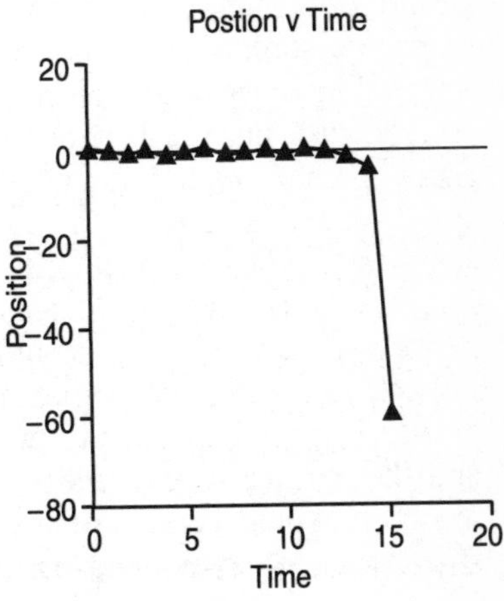

Figure 14A Out of chaos

which could wipe out the human race just as dinosaurs were wiped out about 60 million years ago. Pockets of survivors might be forced to live as our ancestors did, subjected to the perils of nature. The asteroids are very large chunks of rock in orbit about the Sun, mostly between Mars and Jupiter. In addition to the Sun's gravity pulling on an asteroid, Jupiter can affect an asteroid as it is the largest planet of the Solar System. The path of an asteroid therefore depends on its distance from the Sun and its distance from Jupiter. Using chaos theory, astronomers reckon that there is a very small possibility of an asteroid behaving chaotically and colliding with the Earth. No doubt other solar systems have the same problems. Any Earth-like planet in any other solar system would be subject to asteroid impacts from asteroids in its own solar system. The probability of intelligent life elsewhere in the Milky Way galaxy is therefore even more remote if it can be wiped out by asteroid impacts. Perhaps the human race is the only intelligent life form in the Galaxy at the present time. Astronomers and scientists are becoming increasingly aware of the need to watch out for collisions with errant asteroids and other large objects from space. The US Government has set up an observatory to search for and track asteroids. However, advanced warnings of an asteroid impact would not shield the Earth from an impact. An international defense system needs to be set up to safely deflect or destroy any asteroid or other large object from space found to be on course to collide with the Earth. Governments would need to provide the funds, perhaps recharging people in the wealthy nations through a small annual surcharge on home insurance policies. This might seem far-fetched but meteorite impacts are not. A large meteorite impact created the Grand Canyon in Arizona. Such an impact on a densely populated area would devastate the area. An asteroid impact would devastate entire continents and throw millions of tons of dust into the atmosphere. The sky would be dark for many years and the surface would freeze. Food supplies would dwindle and transport would cease. An international defense system seems a small price to pay in comparison.

Science and society

Modern science has provided us with the means of finding out about nature on a scale from quarks to quasars. Many more scientific discoveries undoubtedly lie ahead. The observations and experiments carried out by scientists provide essential guidance for the theories of science. Before the Scientific Age was ushered in by Galileo, few dared to challenge the medieval Church on the theory that the Earth must be at the centre of the Universe. The freedoms we now take for granted such as human rights and democracy would perhaps not have been won if the struggle for intellectual freedom, the hallmark of modern science since Galileo, had been lost. Perhaps the real long-term challenge for the future is to extend the immense benefits of science to the entire human race and to find out how to sustain these benefits indefinitely. Scientific knowledge is vital for a better future. I hope you have enjoyed using this book and that you will take your physics studies further.

Answers to Chapter 1 questions

1. (a) 1.5×10^4 m (b) 4.5×10^{-5} m (c) 6.5×10^8 kg (d) 3.5×10^{-9} kg
2. (a) 7.5×10^{-2} m (b) 1.59 m (c) 5.6×10^7 m (d) 6.5×10^{-2} m (e) 2.7×10^{-8} kg
3. (a) 10^3 (b) 10^6
4. (a) 12500 kg (b) 0.005 m^3 (c) 0.012 m^3
5. (a) 0.001m^3 800 kg / m^3
6. (b) 3.1 kg
7. 72 kg
8. 6000 kg
9. 930 kg / m^3
10. 7500 kg / m^3

Answers to Chapter 2 questions

1. (a) 720 m (b) 4.5×10^5 m (c) 1.8×10^4 m
2. (a) 67s (b) 25s (c) 1667 s = 27 min 47 s
3. (a) 12.5 m/s (b) 10 m/s (c) 133 m/s
4. (a) 13 m/s (b) 310 m/s (c) 28 m/s (d) 200 m/s
5. (a) 72000 m (b) 20 m/s
6. (a) 10 m/s (b) 200 m
7. (a) 0.25 m/s (b) 0.50 m/s (c) 0.0625 m/s^2
8. (a) 60 m/s (b) 2400 m (c) 3.0 m/s^2
9. (a) 15 m/s (b) 11 m
10. (a) 4 m/s^2 (b) 5000 m (d) 7000 m (= 5000 m + an additional 2000 m)

Answers to Chapter 3 questions

2. (a)(i) 19.6 N (ii) 39.2 N
3. (a)(i) 8 kg (ii) 103 N (b) 103 N
4. (a)(i) 8 N (ii) 24 N (b) (i) 5 m/s^2 (ii) 4 m/s^2
5. (a) 2 m/s^2 (b) 1600 N
9. (b) 100 N

Answers to Chapter 4 questions

2. (a) 2000 J (b) 2000 J
3. (a) 900000 J (b) 9000 seconds
4. (a) 392 N (b) 1176 J (c) 47 W
5. (a) 2425 J
6. (a) 1000 J (b) 5000 J (c) 20%
7. (a) 550 kJ (b) 0.55
8. (a) 1000 J (b) 50 W
9. (a) (i) 3000 J (ii) 900 kJ (iii) 40 J (b)(i) 392 J (ii) 24 J (iii) 1250 J
10. (a) 8820 J (b) 3750 J

Answers to Chapter 5 questions

1. (a)(i) 0°C, 273 K (b)(i) 100°C (ii) 373 K (c) –273 °C, 0K
3. (a) No 1: roof = 1200 W, window = 500 W, wall = 1800 W; total = 3500 W
   (b) No 3: roof = 300 W, window = 300 W, wall = 600 W; total = 1200 W
   (c) No 1 = £1.09 No 3 = £0.37
4. (a) 567 kJ (b) 32 kJ (c) 599 kJ
5. (a) 25 °C (b)(i) 21.6 kJ (ii) 12 °C
6. (a) 3.4 MJ (b) 18.8 minutes
7. (a) 1.15 MJ (b) 3.4 kg
8. (a) (i) 840 kJ (ii) 53 kJ (b) 71 W
9. (a) (i) 13 kJ (ii) 67 kJ (b) 67 W
10. (a) 180 kJ (b) 0.078 kg

Answers to Chapter 6 questions

2. (a) 0.54 (b) 0.51
3. 17 000 MW approx (b) 33000 MW approx
5. (a) 6 (b) 2500 (c) 100 km

Answers to Chapter 7 questions

1. negative, gains, loses
2. (a) the coulomb (b)(i) 0.30 C (ii) 9 C

3. (a) 12 hours (b) 36 W
4. (a) 15 V (b) 37.5 W
6. (b)(i) 0.5 A (b) 24 Ω
9. (b)(i) 1; 12 (ii) 1; 36 A, 2; 3 A
10. 5 A
12. (a) £7.75 (b) 96 units, 68%

Answers to Chapter 8 questions

1. 1.20 m
2. See p. 115
3. See Fig. 8A(b)
4. (a) Refraction is the change of direction of a light ray when it passes from one transparent substance or air into another transparent substance. (b) 35°
5. See Fig. 8B(b).
6. (a) Light consists of tiny particles called corpuscles.

   (b) The speed of light in water was found to be less than in air, as predicted by wave theory.
9. See summary p. 133
10. (a) Red (b) 500 million million hertz (c) Infrared radiation, radio waves, microwaves

Answers to Chapter 9 questions

1. (a)(i) negative (ii) positive (b)(i) 16 atomic mass units (ii) +8 *e*, where *e* is the magnitude of the charge of the electron (iii) + *e*
2. (a) molecule, atoms, atom (b) isotope, protons, neutrons (c) proton, electron
4. (a) a covalent bond (b) an ionic bond
6. (a) crystalline (b) polymer (c) crystalline (d) amorphous (e) polymer
7. (a) a good electric insulator (b) It would melt
9. (a) water, cream, syrup, tar (b) To make it less viscous (c) wallpaper paste, custard
12. (a) 19.6 kPa (b) 16 kPa

Answers to Chapter 10 questions

1. (a) particle (b) wave (c) particle
3. (a) AB, AC, BC (b) BC
4. (a) halved (b) more detailed
5. (a) Mass loss = energy loss / $c^2 = 1.1 \times 10^{-36}$ kg which is much less than the mass of the atom
6. 4.4×10^9 kg

Answers to Chapter 11 questions

1. pressure, voltage, ions, ions, ions, ion, ion
2. (a) increased deflection (b) reverse deflection
4. (a) 0.41 per second (e) beta
5. (a) 250 million (b) 62.5 million (c) 238
6. (a) 0 (b) $+2e$ (c) $-e$
7. (a) $-e$ (b) 0 (c) $+e$
8. (a) uud (b) udd
9. d changes to u
10. uuu

Answers to Chapter 12 questions

1. 38 p, 59 n
2. (b) 7.4 MeV
6. approx 1500 kg

Answers to Chapter 13 questions

6. (b) Red shift is proportional to distance
   (c) The Universe is expanding
7. (a) 400 million light years (b) 80000 km/s.
9. Big Bang Quarks and antiquarks Neutrons and protons Atom Galaxies
10. (a)(i) about a minute (ii) 100000 years old

GLOSSARY

Absolute zero The lowest possible temperature (= –273 °C).

Acceleration Rate of change of velocity.

Atom The smallest particle of an element characteristic of the element. Every atom contains a nucleus (which consists of protons and neutrons, except for the hydrogen atom which has a single proton as its nucleus) surrounded by electrons.

Bond Name for any type of force that holds two particles together.

Binding energy Energy needed to separate a nucleus into its constituent neutrons and protons.

Charge There are two types of charge, referred to as positive and negative. Particles that possess the same type of charge repel each other. Particles that possess opposite types of charge attract each other. Charge is quantized in whole number multiples of e, the charge of the electron.

Density Mass per unit volume of a substance.

Diffraction Spreading of waves after passing through a gap or round an obstacle.

Diffusion Gradual spread of randomly moving particles in a substance to a uniform distribution.

Diode Electronic component that allows current to pass through it in one direction only.

Elasticity The physical property of a substance that enables it to regain its shape after being distorted.

Electromagnetic spectrum The spectrum of electromagnetic waves. The electromagnetic spectrum comprises gamma rays and X-rays, ultraviolet radiation, visible light, infrared radiation, microwaves, radio waves. All electromagnetic waves travel at a speed of 300 000 km/s through space.

Electron A negatively charged particle that is in every atom, moving round the nucleus. The charge of the electron, *e*, is equal to 1.6×10^{-19} C.

Electron volt (eV) 1.6×10^{-19} J, defined as the work done when an electron moves through a p.d. of 1 volt. 1 MeV = 1.6×10^{-13} J.

Energy The capacity to change the motion of an object.

Force Any interaction that can change the motion of an object.

Frequency The number of cycles of oscillation of an oscillating object, each cycle being from one extreme to the opposite extreme and back.

Gradient The gradient of a straight line on a graph is the change of the quantity plotted on the *y*-axis / corresponding change of the quantity plotted on the *x*-axis.

Grains Crystalline structure in a metal.

Gravitation There is a force of gravitational attraction between any two masses. The force is inversely proportional to the square of the distance between the centres of the masses.

Half life The time taken for half the number of nuclei of a radioactive isotope to disintegrate.

Interference Where two waves pass through each other, reinforcement occurs where crests meet or where troughs meet. Cancellation occurs where a crest meets a trough.

Internal energy The energy of an object regardless of its state of motion or its position.

Ion A charged atom. An uncharged atom contains equal number of electrons and protons. Removal of an electron makes the atom into a positive ion. Addition of an electron makes the atom into a negative ion.

Isotope The isotopes of an element are forms of the element which each have the same number of protons but a different number of neutrons in each nucleus.

Kinetic energy The energy of an object due to its motion.

Lepton Elementary particles such as the electron and the positron that are not quarks.

Mass A measure of the quantity of matter in an object.

Molecule Two or more atoms joined together.

Momentum The product of the mass and the velocity of an object.

Neutron An uncharged particle slightly heavier than the proton. Every atom contains a nucleus which is composed of one or more protons and neutrons.

Nuclear fission When a large nucleus splits into two fragments.

Nuclear fusion When light nuclei are fused together.

Photon Light is composed of photons. Each photon is a wave packet of electromagnetic energy. The energy of a photon of frequency f is equal to hf, where h is the Planck constant.

Potential energy The energy of an object due to its position.

Power Rate of transfer of energy.

Pressure Force per unit area acting at right angles to a surface.

Proton A positively charged particle which is the nucleus of the lightest atom, the hydrogen atom.

Quarks Particles that combine in threes to form neutrons or protons.

Red shift The increase of the wavelength of light from a receding object due to its receding motion. This increase causes a shift in the line spectrum of the light from the object towards the red part of the spectrum.

Resistance Voltage per unit current needed to make electricity flow.

Scalar A non-directional physical quantity.

Speed of light Distance travelled per second by light. The symbol c denotes the speed of light in free space. $c = 300000$ km/s.

Superconductor An object with zero electrical resistance.

Superfluid A fluid that has zero viscosity.

U-value Heat flow per square metre passing through a wall or panel, etc. when the temperature difference across the object is 1°C.

Vector A physical quantity that has magnitude and direction, e.g. force.

Velocity Rate of change of distance in a given direction.

Viscosity A measure of flow resistance in a fluid.

Voltage Power per unit current delivered between two points in a circuit.

Wavelength The least distance along a wave between two crests.

Weight The force of gravity on an object.

Work Work is done when a force moves its point of application in the direction of the force.

APPENDIX

SPREADSHEET FOR 'OUT OF CHAOS' FIGURE 14A

Note The symbol $ is for absolute cell references

1 Key into cells A1 and A2, the text expressions 'Parameter =' and 'Initial position ='.

2 Key the parameter value into cell B1. Key the initial position into cell B2.

3 Key text headings 'Time', 'Old position' and 'New position' into cells A3, B3 and C3 respectively.

4. Key 0 into cell A4; B2 into cell B4 and the formula 4*B1*D2*1(1–(D2)) into cell C4 for the new position calculated from the old position and the parameter according to the equation
new position = 4 × parameter value × old position × (1 – old position).

5 Key (A4) +1 into cell A5; key C4 into cell B5; copy the contents of C4 into C5.

6 Copy cells A5, B5 and C5 down columns A, B and C.

7 To chart the results, plot col. C on the y-axis against col. A on the x-axis.

INDEX